PROBLEM SOLVING IN ELEMENTARY ALGEBRA

Watson Fulks
University of Colorado

Charles E. Merrill Publishing Company
A Bell & Howell Company
Columbus, Ohio

To Gloria

Copyright © 1970 by Charles E. Merrill Publishing Co., Columbus, Ohio. All rights reserved. No part of this book may be reproduced in any form, electronic or mechanical, including photocopy, recording, or any information storage and retrieval system without permission in writing from the publisher.

International Standard Book Number: 0-675-09343-0

Library of Congress Catalog Card Number: 76-118079

1 2 3 4 5 6 7 8–76 75 74 73 72 71 70

Printed in the United States of America

PREFACE TO THE TEACHER

This booklet is a text supplement which can be used with any of the standard text books on elementary algebra. The reason for its applicability on such disparate levels is that the emphasis is on *method* rather than content. Also, the method is applicable to "word problems" or "stated problems" regardless of the background of the students or the difficulty of the problems. The primary aim is to get the students to read and *reread* each problem until they understand it thoroughly.

The discussion offered in this book is appropriate for several reasons. Anyone who has ever taught a course in algebra has surely found that word problems present probably the most troublesome pedagogical difficulty in that course. Any material which offers some aid in this matter can be useful. Furthermore, some of the more modern texts have tended to de-emphasize such problems, for understandable reasons, but to a greater extent than may be justifiable. The presentation this book provides will enable teachers to increase the emphasis in this area to whatever extent they desire. But the primary aim is improvement of teaching of the subject, not the injection of additional material into the course.

It has become quite common in the languages, the humanities, the social sciences, and to a lesser extent, the sciences, to use supplementary material in courses at all levels, since few if any texts in any area are completely adequate for modern presentations. Teachers in these areas frequently make extensive use of supplementary matter of varying sorts. There seems to be no reason why mathematics teachers should not have these same options available—none, that is, except a shortage of such material.

Teachers who examine this book may see many ways to use it in classroom situations. At the risk of boring the more experienced teacher, a brief description of possible ways to use the book follows. This, in any case, expresses the author's intentions for the material.

I. The teacher can spend from a few days to a week or so covering the material more or less thoroughly with the class.

II. If the class structure permits, some of the better students can present the material to the rest of the class.

III. The better students can act as "teacher's aides" to help the other class members read and apply the material.

IV. The students can use the book completely on their own, without aid.

V. Clearly, many combinations of the previous suggestions are possible. For example, the teacher can spend a day giving explanations and examples, then turn to either II, III, or IV.

Units of classroom work of considerable variations in difficulty can be extracted from the material presented here. For instance, a beginning class could spend time only on the earliest and simplest problems in each of the chapters through Chapter 10. More advanced units could include the harder problems in these chapters. Finally, a unit based on the last three chapters would challenge the best pre-calculus students.

Elementary classes should study Chapter 1 carefully, at least through the first method of solution, and probably through the second. The rest of the chapter is also useful for the elementary classes which have had a little practice, and should be studied immediately by the senior high school classes, and certainly by all college level classes.

Chapter 2 provides some practice in reducing written statements to algebraic expressions. The procedure is to begin with numerical statements and to modify these by replacing some or all of the numbers by letters. This technique is, of course, critical and plays a very important role throughout the rest of the book.

Chapter 3 provides the explanation of the method used in this book, first in outline form and then in great detail. Because of this detail in the explanation of the BASIC OUTLINE, one should use this expanded description primarily as a reference. Thus a reasonable way of covering this material would be to go over the BASIC OUTLINE itself carefully, and read through the detailed explanation once to get the general idea. Then, in handling individual problems, this explanation can be referred to and studied in as much detail as necessary. This gives an immediacy to these explanations which would be absent if great stress were put on it before the student encountered the problems. After these first chapters, a great many options are open as explained earlier.

One point should be emphasized. It is to be hoped that as a student progresses and gains facility in working stated problems, he will be able to do more and more of the work of the outline mentally. It is certainly not necessary for every student to write down all the details on every problem. But for all, except possibly the very best, it is advisable to start that way. The ideas presented in this booklet, as you will readily agree, are used at least implicitly by good students. But even for them, it seems worthwhile to make the procedure explicit, for it gives them a definite, clear approach to rely on when working difficult problems. For all students the outline should provide a great deal of help in what remains one of the thorniest areas in elementary mathematics.

Finally, there are clearly many ways in which problems could be classified. The one used here is a convenient one, but by no means the only reasonable one. And certainly no claim is made that the classification is exhaustive. (For example, lever problems are not included.) But the coverage is sufficiently broad that any other types of problems can easily be treated by the general method used here.

I would like to take this opportunity to thank Earl and Elaine Hasz who were very helpful in the selection of the examples and exercises used in this book.

TABLE OF CONTENTS

Chapter 1	General Remarks and a Solved Example	1
Chapter 2	Symbolic Expressions	7
Chapter 3	The Basic Outline	11
Chapter 4	Age Problems	17
Chapter 5	Number and Money Problems	27
Chapter 6	Geometry Problems	33
Chapter 7	Work Problems	41
Chapter 8	Mixture and Value Problems	47
Chapter 9	Interest Problems	55
Chapter 10	Rate Problems	61
Chapter 11	Maximum and Minimum Problems	73
Chapter 12	Acceleration Problems	79
Answers		87

1
GENERAL REMARKS AND A SOLVED EXAMPLE

If you are like most people you find word problems more difficult than formal exercises. The basic reason is, I think, that they demand that you yourself contribute something more than a straightforward application of the rules of algebra which you have learned. They present a need for a clear analysis of assorted facts and for the translating of these facts into mathematical language—a need not present in mechanical exercises. This makes them more difficult, but at the same time more challenging. The purpose of this little booklet is to present a procedure which will help you analyze such problems.

But you must not expect too much—the methods for the solution of problems of this sort can never be reduced to a turn-the-crank-and-grind-out-the-answer process. That is, no one can say, do first "this," then do "that" and finally do "this," and the answer will come out. The best that can ever be done, and that is what we attempt here, is to say, "do this sort of thing," and "that sort of thing," and you have a *reasonable approach* to word problems. What we will do is give a *method of attack* on problems, which amounts to saying how one should go about analyzing the problem and proceeding with its solution.

If you follow carefully and faithfully the directions given here, you can learn to analyze word problems *systematically* so as to reduce the difficulty. But, and we will repeat this warning often, such analysis will not remove all difficulties. As Aristotle is supposed to have said to Alexander the Great when he was tutoring Alex in the mysteries of geometry, there is no royal road to mathematics. But by learning to approach problems systematically you can learn to get quickly to the heart of a problem, to know exactly where the *real* difficulty lies. This, it is important to realize, is already a big step toward understanding what to do.

General Remarks and a Solved Example

But, let me remind you of something else: in this, as in most other work you attempt, your results will be in proportion to the amount of hard work and real interest you can bring to bear on the problems. The frame of mind in which you approach the problems, at least the more difficult ones, has a distinct influence on your chances of success in applying the methods we suggest here. First, the attitude you should definitely *not* have, is that this is a very disagreeable task and because you feel obliged to do so, you will put in some time staring at the statement of the problem. This, I am sure you can see, is not likely to lead to the solution of a problem of more than routine difficulty. The more genuine interest you have the better your chances of success will be, because you will be willing to work harder. If you find a problem intriguing, as a puzzle can sometimes be, so that you are anxious to find out how it can be solved, then you are in the best frame of mind in which to approach a stated problem. It is of course true, and there's no use pretending otherwise, that not all problems generate such a level of interest. But a sprinkling of interesting ones among others which are not quite so enjoyable can sustain the interest of anyone who is willing to give it a good try.

We will now take an example of a word problem in algebra and work carefully through it in great detail. We will take a fairly simple one, one which requires a typical method of solution. Then later, we will review the general procedure we used and, in Chapter 3, set up an outline for attacking all word problems.

The first thing to do is to *read the problem carefully:*

> A wholesale oil dealer sold 10,000 gallons of lubricating oil for $15,250. Some of the oil sold at $2.50 a gallon, and the rest at $1.00 a gallon. How much of each kind of oil was sold?

Again, let me emphasize that it is necessary to *read the problem carefully*, noting and understanding all the parts.

Now, let us ask ourselves, What is the unknown or unknowns? That is, what is wanted in the solution of the problem? In this problem we want to know how much (that is, how many gallons) of each kind of oil was sold.

Next we observe that we were given the following facts:
- (a) The total amount of oil was 10,000 gallons.
- (b) The total value of the oil was $15,250.
- (c) Some oil sold at $2.50 per gallon.
- (d) The rest sold at $1.00 per gallon.

General Remarks and a Solved Example 3

The next step is to choose *a symbol* to represent the unknown:

Let x be the *number* of gallons of oil that sold at $2.50 per gallon.

Then $10,000 - x$ is the *number* of gallons of the oil which sold for $1.00 per gallon.

Note that we have used here the information that the total amount of the oil was 10,000 gallons. But we have not made any use of our knowledge of the prices of the oil. To use this we will now express the value of each kind of oil in terms of our unknown, x. Clearly the price of each kind of oil is the number of gallons times the value per gallon:

$2.50x = 2\frac{1}{2}x = \frac{5}{2}x =$ the total value (that is, the number of dollars worth) of $2.50 oil

$1.00(10,000 - x) = 1(10,000 - x) = 10,000 - x =$ the total value (in dollars) of the $1.00 oil

We have now used our information about the total amount (gallons) and the value (dollars) of each kind of oil. The only piece of information left which we have not used is that the total value of all of the oil is $15,250. But at this stage it should be clear that the way in which we use this is to observe that the whole is the sum of its parts. In our case this means that the total value of the oil is the sum of the value of the $2.50 oil and the value of the $1.00 oil:

$$\underset{\substack{\text{Value of} \\ \$2.50 \text{ oil}}}{\frac{5}{2}x} + \underset{\substack{\text{Value of} \\ \$1.00 \text{ oil}}}{(10,000 - x)} = \underset{\substack{\text{Total} \\ \text{Value}}}{15,250} \qquad (1)$$

This gives us an *equation* concerning the unknown number x, that is, concerning the number of gallons of the $2.50 oil. We can easily solve this equation for x. Multiplying both sides by 2, we get

$5x + 20,000 - 2x = 30,500$
$3x = 10,500$
$x = 3,500 =$ the number of gallons of the $2.50 oil
$10,000 - x = 10,000 - 3,500 = 6,500 =$ the number of gallons of the $1.00 oil.

We easily check our solution:

Value of $2.50 oil $= 3,500 \times 2.50 = 8,750$
Value of $1.00 oil $= 6,500 \times 1 = 6,500$
Total value $ = 15,250$

4   General Remarks and a Solved Example

Let us review the solution of this simple example, or, more properly speaking, let us review the *method* we used in its solution.

1. We read the problem carefully.
2. We made very clear in our minds what the problem asked us to find. (What is the unknown?)
3. We listed *all* the given information.
4. We selected a symbol to represent the unknown number, and *in terms of this symbol* we expressed the given information.
5. We wrote down the equation expressing a relationship between the given numbers and the symbolized numbers. (The whole is equal to the sum of its parts.)
6. We solved the equation.
7. We checked the solution.

We might ask ourselves if this is the only way we could set the problem up and find the solution. In its main features the answer is, essentially, yes, though we might not have bothered to go into such great detail. But in the selection of the symbol for the unknown and in the details we could have proceeded in other ways. We now look at two of the other ways we could have solved the problem.

If we had set $y =$ amount of the \$1.00 oil, that is, $y =$ the number of gallons of the \$1.00 oil, then the amount (gallons) of the \$2.50 oil would have been $10{,}000 - y$. The values of the two different kinds of oil would have been

$$y = \text{value (in dollars) of the \$1.00 oil}$$

$$2.50(10{,}000 - y) = \left(\frac{5}{2}\right)(10{,}000 - y) = \text{value (in dollars) of the \$2.50 oil}$$

From this we then get a different equation:

$$y + \left(\frac{5}{2}\right)(10{,}000 - y) = 15{,}250 \qquad (2)$$

Now this equation is different, but for our word problem it *should be equivalent*. That is, the final result that we get should be the same as before. Hence we should get $y = 6{,}500$ from equation (2). (Do *you* get this?)

General Remarks and a Solved Example 5

Our next alternative solution is a combination of the first two. If we note that there are really two unknowns, namely, the amounts (number of gallons) of the two kinds of oil, we can use different letters for these two unknowns. So we let

$$x = \text{gallons of the \$2.50 oil}$$
$$y = \text{gallons of the \$1.00 oil}$$

Then

$$2.50x = \left(\frac{5}{2}\right)x \text{ dollars} = \text{value (dollars) of the \$2.50 oil}$$

$$1.00y = y \text{ dollars} = \text{value (dollars) of the \$1.00 oil}$$

This uses only parts (c) and (d) in our list of information. Since the whole is the sum of its parts, part (a) implies that

$$x + y = 10{,}000 \tag{3}$$

and part (b) implies that

$$\left(\frac{5}{2}\right)x + y = 15{,}250 \tag{4}$$

These two equations can be solved for x and y to get $x = 3500$ and $y = 6500$. The connection between this last method and the first two can be seen by examining equation (3). If we solve this equation for y to get $y = 10{,}000 - x$, and substitute this value of y into equation (4), then (4) reduces back to equation (1) in our first method. However, if we solve (3) for x to get $x = 10{,}000 - y$ and substitute this into (4), then this time, (4) reduces to equation (2) of the second method.

The general outline that we wrote down in the review of our first solution of our example would serve very well for solving most word problems. As you will see in Chapter 3 we use a modification of this outline for our basic set of rules for attacking all stated problems. The modifications to be made will consist largely of amplifications, and the addition of one more item to cover a situation which did not arise in this chapter.

2
SYMBOLIC EXPRESSIONS

A technique which must be understood in working stated problems is that of expressing information in terms of unknown numbers. For instance, in the first solution of our example in Chapter 1, we chose the symbol (letter) x to represent the number of gallons of the $2.50 oil and got

$$10{,}000 - x = \text{the number of gallons of the \$1.00 oil}$$

$$\frac{5}{2}x = \text{value (dollars worth) of the \$2.50 oil}$$

and

$$10{,}000 - x = \text{value (dollars worth) of the \$1.00 oil}$$

It is essential (see Steps 3 and 6 of the BASIC OUTLINE) that the student learn this technique, which is very hard to teach. This is because it requires the student to *read* and understand the relationships stated in the problem. Apparently, as with the broader aspects of stated problems as well as many other things, students learn this best by practice, starting with simple cases and progressing to more difficult ones.

Since many algebra texts give some attention to this, and since the student gets still more practice in actually working the problems, we give here only a few such examples.

Write down the following in numerical or symbolic (algebraic) form.

1. A number twice as large as 3.
2. A number twice as large as x.
3. A number four less than three times five.
4. A number four less than three times x.
5. A number y less than three times x.
6. A number y less than z times x.
7. Sixty percent of 10, of 4.
8. Sixty percent of x.

Symbolic Expressions

9. Sixty percent of $(x - y)$.
10. The distance travelled by a car going 50 miles an hour for four hours.
11. The distance travelled by a car going x miles an hour for four hours.
12. The distance travelled by a car going 50 miles per hour for y hours.
13. The distance travelled by a car going x miles per hour for y hours.
14. The time (hours) required for a car to go 200 miles at 40 miles per hour.
15. The time required for a car to go 200 miles at x miles per hour.
16. The time required for a car to go y miles at 40 miles per hour.
17. The time required for a car to go y miles at x miles per hour.

Answer the following in numerical or symbolic form.

18. If John is 10 years old, what was his age 3 years ago?
19. If John is x years old, what was his age 3 years ago?
20. If John is 10 years old, what was his age y years ago?
21. If John is x years old, what was his age y years ago?
22. If Jean is 23 years old and Jane is 21 years old, how old will Jean be when Jane is as old as Jean is now?
23. If Jean is y years old and Jane is 21 years old, how old will Jean be when Jane is as old as Jean is now?
24. If Jean is 23 years old and Jane is z years old, how old will Jean be when Jane is as old as Jean is now?
25. If Jean is y years old and Jane is z years old, how old will Jean be when Jane is as old as Jean is now?
26. If Tom is 15 years old, what is ⅓ of his age?
27. If Tom is x years old, what is ⅓ of his age?
28. What is the value, in cents, of 3 quarters? of 5 dimes? of 3 quarters and 5 dimes?
29. What is the value in cents of x quarters? of y dimes? of x quarters and y dimes?

Symbolic Expressions

30. If John can paint a wall in 5 hours, what fractional part of it can he paint in one hour? in two hours? in x hours? in 5 hours?
31. If John can paint a wall in x hours, what fractional part of it can he paint in one hour? in two hours? in x hours?
32. If John can paint a wall in 8 hours and Tom can paint the same wall in 4 hours, what part can they paint together in 1 hour? in x hours?
33. If John can paint a wall in x hours, and Tom can paint the same wall in y hours, what part of it can they paint together in 1 hour? in 2 hours? in z hours?

3
THE BASIC OUTLINE

The outline for working word problems discussed here is one which you should use for reference in working *all* word problems. Follow this outline and you will greatly simplify the solutions of many problems. But this is *not* a promise of miraculous results. These, if they come, must come from you. The most that a teacher, or a book, or both, can do to help in solving stated problems is to help you organize your material systematically and to point out the difference between (a) that which is difficult because it is simply a conglomeration of mixed up things which most everyone, by the use of care and the proper routine, can learn to analyze and untangle, and (b) that which is intrinsically hard and requires insight to understand.

If you will follow this outline faithfully, you can learn to do the routine things well, and then you will have learned to put the difficulty where it belongs, namely, on the intrinsically hard part of a problem. This, as was said earlier, is a big step toward the solution of the most difficult problems.

We now state the BASIC OUTLINE and follow it by a more detailed discussion of the meaning of the abbreviated statements of which it is composed.

BASIC OUTLINE

1. Read the problem carefully, *several times* if necessary, until you can answer all questions and do all the steps which follow.

2. What is asked for? What is the unknown? Write this out in words.

3. Find all the given data. What is the given information? Write this out in words.

The Basic Outline

4. Decide if there is any information that is supposed to be common knowledge which will be useful to you in working the problem. If so, write it out in words.

5. *Select letters to represent the unknowns.*

6. Express the given information *in terms of the unknowns.* Drawing a figure may be helpful.

7. Set up the equation or equations using the symbolic language developed in Step 6. Again a figure may be very useful.

8. Solve the equation or equations.

9. Check the solution against the *original* statement of the problem.

This BASIC OUTLINE should be looked upon as just that—an outline. We now proceed to a more complete examination of the procedure outlined here in which we state more carefully and more fully what should be done at each step in following this outline. This section, including the BASIC OUTLINE and the following discussion should be read over carefully to fix it firmly in your mind. You should then refer back to it constantly while working the problems, so that by the time you have worked through the book these pages should show much more wear than the rest of the book.

Of course, as you go along, you will develop greater and greater facility in analyzing problems so that more and more of the work can be done mentally. But at the beginning, and when you have difficulty with hard problems later in the book, you should follow the outline in great detail.

Now let us go through the outline examining precisely what is meant by each item, and what should be done at each step.

1. Read the problem carefully. Read it several times if necessary, paying close attention to every word. The statements of most problems are short, but they usually contain a great deal of information in very compact form, some of it subtle, some of it disguised, and sometimes useful information may not be stated explicitly, but is simply assumed to be common knowledge that you are already familiar with. (See Step 4 in the BASIC OUTLINE.)

The difference between understanding a problem and not understanding it may sometimes hinge on the correct reading and understanding of a single word. To make sure that you are reading every word, and *absorbing every word*, it may be helpful to copy the problem slowly, thinking of the meaning as you write it down.

To sum up: *read carefully.*

2. What is asked for? What is the unknown or unknowns? Write out the answer to this question in words, and be careful to be precise. What I mean by "be precise" is perhaps explained best by some examples. Thus the unknown may be "the speed of the second car," but it will never be "the second car." Or it may be "the length of a side of a square," but not "the square." Remember, *the unknown will be a number.* In our example in Chapter 1 the unknowns were the *number* of gallons of certain kinds of oil.

3. Find *all* the given data. Write this down carefully, remembering that this information will be expressed in *numbers.* Again, as in Step 2, be careful to be precise. For example: "the speed of the car is 33 miles per hour," *not* "the first car is 33."

Of course, in the broad sense, the answer to the question of what is given includes *everything* stated in the problem. But here we mean to separate all this out into individual pieces of information as we did with our list (a), (b), (c), and (d) on page 2 in the example in Chapter 1. There may be different ways to write this down, but the important point here is to *be sure you have it all.*

4. Is there any common knowledge which will be useful to you in working this problem?

Suppose you are examining a problem about the ages of a father and son. It might be useful to realize that the father's age is greater than the son's age. This is, of course, a fact that everyone knows, but not everyone may *remember to use* this information at the right time.

As another example, problems are sometimes stated about triangles. Now it might well be that the author of the problem expects you to remember and use the fact that the sum of the angles of a triangle is 180°, without his supplying this information in the statement of the problem. Everyone who has studied geometry knows this. But, will *you* recall this useful bit of information when you need it? If a problem is stated about a clock it may be important to know that the minute hand turns at a rate of 6° per minute. How fast does the hour hand turn?

The point of these examples is to illustrate the fact that if a problem is about a familiar object or situation, you may be expected to supply some useful information out of your general knowledge.

5. Select letters to represent the unknowns. Remember that the unknown is a *number*. For example, you should write

$$\text{let} \quad x = \text{John's age}$$

and not

$$\text{let} \quad x = \text{John}$$

Or

$$\text{let} \quad x = \text{length (for example, in feet) of a side of the square}$$

not

$$\text{let} \quad x = \text{the square}$$

Or finally

$$\text{let} \quad x = \text{the speed (in miles per hour, say) of the second car}$$

not

$$\text{let} \quad x = \text{the second car}$$

It is true that usually, though unfortunately not always, students understand that the unknown letter stands, for example, for the speed of the car, and not for the car itself. Such sloppiness as writing "$x = $ the car," rather than "$x = $ the speed of the car" causes no real trouble in simple cases, but the looseness of thought and lack of precision it indicates is what *can* cause trouble in harder and more difficult situations. So get in the habit of stating clearly and accurately exactly what you mean by the symbols (i.e., letters) you use, for these symbols have no *intrinsic* meaning of their own. They mean what *you* say they mean, and you cannot use them properly and effectively unless you yourself have a very *clear and precise* understanding of just what you do want them to mean.

As you know, and as you have seen in the example we discussed in Chapter 1, we may have a choice of two or more unknowns which we can choose to represent by a letter. There is not always one particular choice which is preferable to the others. Thus in the example in Chapter 1, it makes no discernible difference whether we choose $x = $ number of gallons of \$2.50 oil or $y = $ number of gallons of \$1.00 oil. And it should be clear that it does not make any difference, from the standpoint of the logic involved, which of the alternative choices we make, provided we express the information correctly and write equations correctly in terms of the unknown we do choose. But it can sometimes make a computational difference. That is, the work of expressing the information in terms of the unknown, setting up the equation, and solving the equation may be simpler with one choice than with another.

The Basic Outline 15

To take a very simple example, a problem may state that one has twice as many dimes as nickels. In this case we could, quite clearly, make the choice

$$x = \text{the number of nickels}$$

then

$$2x = \text{the number of dimes}$$

or we could write

$$x = \text{the number of dimes}$$

then

$$\frac{1}{2}x = \text{the number of nickels}$$

6. Express the given information in terms of the unknowns. Here is where one is most likely to notice the difference, if indeed there is one, made by the choice of the unknown in Step 5. If the expressions resulting from one choice are noticeably simpler than those from another choice, it is quite likely that the rest of the problem will also be easier if we use the simpler expressions. In the actual working out of a problem, a good rule of thumb to follow is that if the expressions you get with one choice look pretty complicated, try another. If they look fairly simple, go ahead and work with the choice you have made. Again, let me say that from the standpoint of the logic involved it does not make one bit of difference which way a problem is set up.

This step is one of the two most critical steps in the BASIC OUTLINE. If a real difficulty is going to appear it is most likely to be here or in Step 7. However, it is more difficult to give general directions to overcome a difficulty here. Certainly you should *always* draw a figure illustrating the problem if you can. And if you are having trouble here, you should re-read the problem and work through all the previous steps to make sure you have done your work correctly so far.

7. Set up the equation or equations from the mathematical expressions you have developed in Step 6, and the information in the previous steps. *Here we are considering the heart of the problem.* Usually the previous steps in the BASIC OUTLINE are easier than this. Thus we may consider that the procedure we have described has been designed to simplify the other steps so that at this point you will have as clear a picture as possible of what is involved. You can then give your full attention to this central part of the problem.

The Basic Outline

There are many different ways to arrive at an equation, depending on the type of problem. It is possible to give some general suggestions, but these will not cover all possibilities. Certainly the *first* thing to do if you can, and if you have not done so already in Step 6, is to *draw a figure*.

One of the general principles used in setting up equations is that the whole is equal to the sum of its parts. We used this in the example in Chapter 1, when we set up our equation. There it took the form that the sum of the dollar values of two kinds of oil was equal to the total value of all the oil. This principle is also used in rate problems in the form that the total distance is equal to the sum of separate distances, or that the total time is the sum of separate times.

Another useful principle is that an equation must be balanced. The same quantity computed in two different ways must still be the same.

8. Solve the equation (or equations). The actual solutions of the equations, once they have been set up, usually give very little trouble, partly because the equations one gets from stated problems are relatively simple.

9. Check the solution against the *original* statement of the problem. The reason for the emphasis on using the original statement for checking should be clear. It is simply that if you check your solution against only the equation you set up, then you are doing just that: checking the solution of the *equation*. But if the equation is wrong, then its solution cannot be expected to be a solution to the problem. Thus you should always check the original statement of the problem.

In working a particular problem, when we come to the end of Step 7 we have clearly completed the *analysis* of the problem. But for the sake of completeness we have added the two final steps. Certainly, the solution should not be considered complete until the problem has been checked.

4
AGE PROBLEMS

Almost all age problems met in elementary algebra concern relationships of the ages of people at certain times. The best way to deal with such problems is to build a table at Step 6 of the BASIC OUTLINE expressing the ages of the people at the different times in terms of the unknown or unknowns. Then Step 7 is relatively easy, for one can use this table to write out equations expressing the relationships given in the problem.

EXAMPLE 1 John is twice as old as his sister Joan. Four years ago he was three times as old. What are their ages?

Solution

1. READ THE PROBLEM.

2. The ages of John and Joan are asked for.

3. (a) *Now:* John's age is twice Joan's age
   (b) *Four years ago:* John's age was three times Joan's age.

4. Nothing here.

5. Let
$$x = \text{Joan's age now}$$
$$2x = \text{John's age now}$$

6.

	Now	Four years ago
Joan's age	x	$x - 4$
John's age	$2x$	$2x - 4$

7. Note that the fact that John is twice as old as Joan has already been used. It was built in at Step 5. The only piece of unused information now left is that four years

18   Age Problems

ago John was three times as old as Joan [3(b)]. From this we get:

$$2x - 4 = 3(x - 4)$$
8.
$$2x - 4 = 3x - 12$$
$$x = 8 = \text{Joan's age}$$
$$2x = 16 = \text{John's age}$$

9. Clearly 16 is twice 8, and

$$4 = \text{Joan's age four years ago}$$
$$12 = \text{John's age four years ago}$$

Since 12 is three times 4 we check the solution.

EXERCISE 1   Amy's father Bill is three times as old as Amy. In thirty-six years Bill will be 1½ times as old as Amy. Find their ages.*

EXAMPLE 2   A year ago Arnold was five times as old as his son Bert is now. Five years ago Arnold was three times as old as Bert will be in two years. How old are they?

Solution
1. READ THE PROBLEM.
2. The present ages of Arnold and Bert are asked for.
3. (a) One year ago Arnold was five times as old as Bert is now.
   (b) Five years ago Arnold was three times as old as Bert will be in two years.
4. Nothing here.
5. Let

$$x = \text{Bert's age now}$$
$$5x = \text{Arnold's age one year ago}$$

---

*Answers to all Exercise problems may be found at the back of the book, pp. 87–88.

Age Problems 19

6.

	5 yr ago	1 yr ago	Now	2 yr hence
Arnold's age	$5x - 4$	$5x$	$5x + 1$	
Bert's age			x	$x + 2$

Notice that in the table we filled in only those items of concern to us in the solution of the problem. If we should need the others later they can easily be supplied.

7. Again 3(a) has been built in in Step 5.
From 3(b) we get

$$5x - 4 = 3(x + 2)$$

8.
$$5x - 4 = 3x + 6$$
$$2x = 10$$
$$x = 5 = \text{Bert's age now}$$
$$5x + 1 = 26 = \text{Arnold's age now}$$

9. These clearly satisfy 3(a). To see that 3(b) is satisfied, we observe that

Arnold's age five years ago $= 21$
Bert's age in two years $\quad = 7$
$21 = 3 \times 7.$

Let us solve this another way. We follow the previous solution up to Step 5:

5. Let
$$x = \text{Bert's age now}$$
$$y = \text{Arnold's age now}$$

	5 yr ago	1 yr ago	Now	2 yr hence
Arnold's age	$y - 5$	$y - 1$	y	
Bert's age			x	$x + 2$

20   Age Problems

7. This time 3(a) has not been built in, so it gives:
$$y - 1 = 5x \qquad (1)$$
And 3(b) gives
$$y - 5 = 3(x + 2) \qquad (2)$$
so we have two equations in two unknowns.

8. If we solve for y in equation (1) we get
$$y = 5x + 1$$
and substituting into equation (2) gives
$$5x + 1 - 5 = 3(x + 2)$$
or
$$5x - 4 = 3(x + 2)$$
which is exactly the equation we got in our first solution, and so this solution now proceeds as did the first one.

EXERCISE 2   In a year, three times Tenny's age will be four less than Martha's age now. And Martha's age in three years will be eight times what Tenny's age was five years ago. Find their ages.

EXAMPLE 3   A young lady has two nieces. Her age is three times the sum of her nieces' ages. Two years from now her age will be five times the age of the younger niece, and it will also be twice the sum of the nieces' ages. Find their ages.

Solution

1. READ THE PROBLEM.
2. We seek the ages of the three people.
3. (a) The young lady's age is three times the sum of her nieces' ages.
   (b) In two years her age will be five times the age of her younger niece.
   (c) In two years her age will also be twice the sum of her nieces' ages.

Age Problems

4. Nothing here.

5. Let
$$x = \text{the young lady's age}$$
$$y = \text{the younger niece's age}$$
$$z = \text{the older niece's age}$$

6.

	Now	2 yr hence
Young lady's age	x	$x + 2$
Young niece's age	y	$y + 2$
Older niece's age	z	$z + 2$

7. From 3(a):
$$x = 3(y + z) \qquad (1)$$
from 3(b):
$$x + 2 = 5(y + 2) \qquad (2)$$
and from 3(c):
$$x + 2 = 2(y + 2 + z + 2) = 2(x + z) + 8$$
or
$$x + 2 = 2(y + z) + 12 \qquad (3)$$

8. Subtracting equation (1) from equation (3), we get
$$2 = -(y + z) + 8$$
$$y + z = 6 \qquad (4)$$

Substituting (4) into (1) gives
$$x = 3(6) = 18 = \text{young lady's age}$$

Using this in (2) we get
$$20 = 5(y + 2)$$
$$4 = y + 2$$
$$y = 2 = \text{the age of the younger niece}$$

Then substituting the computed values for x and y back into (1) gives
$$18 = 3(2 + z)$$
$$6 = 2 + z$$
$$z = 4 = \text{the age of the older niece}$$

Age Problems

9. We verify 3(a):

$$18 = 3(2 + 4) = 3 \times 6 = 18$$

We verify 3(b): Then the young lady will be 20 and her younger niece 4. $20 = 5 \times 4$. We verify 3(c):

$$20 = 2(6 + 4) = 2 \times 10$$

EXERCISE 3 John is as much older than Juan as Juan is older than Sean, and six years ago John was twice as old as Sean. In ten years the sum of John's and Sean's ages will be five times Juan's age five years ago. Find their ages.

EXAMPLE 4 Three young ladies are comparing ages. They find that the sum of Jean's and Joan's ages now is ten more than twice Jane's age a year ago. Fifteen years ago Joan was twice as old as Jane, and two years ago Joan was as old as Jean will be when Jane is as old as Jean is now. How old is each?

Solution

1. READ THE PROBLEM.

2. The present ages of the three young ladies is sought.

3. (a) The sum of Jean's age and Joan's age is 10 plus two times Jane's age a year ago.
   (b) Fifteen years ago Joan was twice as old as Jane.
   (c) Joan's age two years ago is the same as Jean's age will be when Jane is as old as Jean is now.

4. Nothing here.

5. Let

$$x = \text{Joan's age now}$$
$$y = \text{Jean's age now}$$
$$z = \text{Jane's age now}$$

Age Problems 23

6.

	15 yr ago	2 yr ago	1 yr ago	Now	Jane as old as Jean now
Joan's age	$x-15$	$x-2$		x	
Jean's age				y	$2y-z$
Jane's age	$z-15$		$z-1$	z	y

The only question here is: What will be Jean's age when Jane is as old as Jean is now? The difference of their ages now is $y - z$, so each of them is $(y - z)$ years older when Jane is y years old. So Jean's age then is her present age plus $y - z$:

$$y + (y - z) = 2y - z$$

7. By 3(a) we get

$$x + y = 2(z - 1) + 10 \qquad (1)$$

by 3(b) we get

$$x - 15 = 2(z - 15) \qquad (2)$$

and by 4(b) we get

$$x - 2 = 2y - z \qquad (3)$$

These equations numbered (1), (2), and (3) can be solved for the ages.

8. The equations become, respectively,

$$x + y - 2z = 8 \qquad (1)$$
$$x - 2z = -15 \qquad (2)$$
$$x - 2y + z = 2 \qquad (3)$$

Subtracting equation (2) from (1) we get

$$y = 23 = \text{Jean's age now}$$

Substituting $y = 23$ into (3) [why not (1)?], we get

$$x + z = 44 \qquad (4)$$

and subtracting (2) from this we get

$$3z = 69$$

or

$$z = 23 = \text{Jean's age now} \qquad (5)$$

Then from (1), (2), or (3), we find, using (4) and (5), that
$$x = 27 = \text{Joan's age now}$$

9. The easiest way to check is to again fill in the table, this time with the numbers:

	15 yr ago	2 yr ago	1 yr ago	Now	Jane as old as Jean now
Joan's age	12	25		27	
Jean's age				23	25
Jane's age	6		20	21	23

3(a)
$$27 + 23 = 2 \times 20 + 10$$
$$50 = 50$$

3(b)
$$12 = 2 \times 6$$

3(c)
$$25 = 25$$

EXERCISE 4 In two years Tim will be as old as the average of Tom's and Tam's present ages. Six years ago Tim's age was twice Tom's age. When Tom is as old as the average of Tim's and Tam's present ages, Tim will be three quarters as old as Tam was the year before.

ADDITIONAL EXERCISES

5. In a year a girl's age will be one-fifth of her mother's age. Two years ago it was two less than one-seventh of her mother's age then. Find their ages.

6. The square of Henry's age is twenty-four less than ten times his age. How old is Henry?

7. A man spent one-fifth of his life as a child. After another tenth of his life he married and three years later a son was born. The son lived ten years longer than his father did, and the man died thirty-four

years before his son died. How long did the man live? (A table is not of much use here.)

8. Amos has two sons, Ben and Charles. Three years ago Amos's age was three times the sum of Ben's and Charles's ages now. When Ben is as old as Charles, Amos will be two years younger than three times the sum of the ages of Ben and Charles. Two years ago Charles was 1½ times as old as Ben is now. Find their ages.

5
NUMBER AND MONEY PROBLEMS

It is fairly common to formulate some word problems about numbers by stating relationships which may exist among their digits. To understand these problems it is necessary to have a clear understanding of the meaning of the digits.

In our decimal system of writing numbers down we use exactly ten distinct symbols: 0, 1, 2, 3, 4, 5, 6, 7, 8, 9. These are called digits. The digits themselves represent the numbers zero through nine. In writing numbers ten through 99 we use two digits. For example, 53 represents the number 53, that is, five tens plus three. Thus the first of these two digits is called the tens' digit, for it tells how many tens must be taken, and the second is called the units' digit. Thus

$$53 = (5 \times 10) + 3$$

If one has a two digit number with the first digit t and the second u then the number is

$$(t \times 10) + u = 10t + u$$

Similarly the first digit of a three digit number is the hundreds' digit, so if the digits of a three digit number are h, t, and u, in that order, then the number is

$$(h \times 100) + (t \times 10) + u = 100h + 10t + u$$

This chapter includes certain money or coin problems because they are handled basically the same way. For example, if in a collection of pocket change there are q quarters, d dimes, n nickels, and p pennies, the total amount of money (in cents) is

$$25q + 10d + 5n + p$$

Also some miscellaneous number problems are included.

EXAMPLE 1 A two-digit number is seven times the sum of the digits, and the tens' digit is two more than the units' digit. Find the number.

28 Number and Money Problems

Solution

1. READ THE PROBLEM.

2. A number is to be found.

3. (a) The number is seven times the sum of the digits.
   (b) The ten's digit is two more than the sum of the digits.

4. The relation of the digits to the number which was explained in the second paragraph of this chapter is assumed to be known.

5. Let
$$x = \text{the units' digit}$$
$$x + 2 = \text{the tens' digit}$$

6. The number is then
$$10(x + 2) + x = 11x + 20$$
and the sum of digits is
$$(x + 2) + x = 2x + 2$$

7. The condition 3(b) was built in at Step 5, and from 3(a) we get
$$11x + 20 = 7(2x + 2)$$

8.
$$3x = 6$$
$$x = 4 = \text{tens' digit}$$
$$x + 2 = 2 = \text{units' digit}$$
so
$$42 = \text{the number required}$$

9. 3(a) $\quad 4 = 7(4 + 2) = 7 \times 6$
   3(b) $\quad 42 = 2 + 2$

EXERCISE 1 The sum of the digits of a two-digit number is 8. If the digits are reversed the new number so formed is 36 more than the original number. Find the number. (Note: By reversing the digits we mean that the units' digit of the old number becomes the tens' digit of the new and vice versa. Thus 38 would become 83, or 23 become 32. The same sort of thing holds for numbers with more than two digits: 528 becomes 825, 3267 becomes 7623, etc.)

EXAMPLE 2 In a three-digit number the sum of the digits is eleven, and the tens' digit is two more than the units' digit. If the digits are reversed, the new number is smaller by 594 than the original number. Find the number.

Solution
1. READ THE PROBLEM.
2. We are asked to find a number.
3. (a) The sum of the digits is 11.
   (b) The tens' digit is two more than the units' digit.
   (c) The new number formed by reversing the digits is 594 smaller than the original number.
4. Nothing here that was not covered in the first few paragraphs in this chapter and the note just preceding the statement of this problem.
5. Let
$$u = \text{the units' digit}$$
$$t = \text{the tens' digit}$$
$$h = \text{the hundreds' digit}$$
6. The sum of the digits is
$$h + t + u$$

The original number is
$$100h + 10t + u$$

and the new number is
$$100u + 10t + h$$

and their difference is
$$99h - 99u$$

30   Number and Money Problems

7. From 3(a) we get
$$h + t + u = 11 \qquad (1)$$
from 3(b) we get
$$t = u + 2 \qquad (2)$$
and from 3(c) we get
$$99h - 99u = 594$$
We can divide this last equation by 99 to get
$$h - u = 6 \qquad (3)$$

8. Substituting $t = u + 2$ into equation (1) gives
$$h + 2u = 9 \qquad (4)$$
and subtracting equation (3) from equation (4) gives
$$3u = 3$$
$$u = 1 = \text{the units' digit}$$
$$t = u + 2 = 1 + 2 = 3 = \text{the tens' digit}$$
and from (3)
$$h = u + 6 = 1 + 6 = 7 = \text{the hundreds' digit}$$
So the number is 731.

9.
$$7 + 3 + 1 = 11$$
$$3 = 1 + 2$$
$$731 - 137 = 594$$

EXERCISE 2   In a three-digit number the units' digit is 4 more than the hundreds' digit, and the sum of the units' and the tens' digits is eleven. If the tens' and hundreds' digits are interchanged, the new number is greater by 450 than the original number. Find the number.

EXAMPLE 3   Arnold has five dollars in nickels, dimes, and quarters. He has twice as many nickels as quarters. How many of each kind of coin does he have, if he has forty-five coins in all?

Number and Money Problems 31

Solution

1. **READ THE PROBLEM.**

2. The number of each kind of coin is asked for.

3. (a) The total value is $5.00 = 500¢.
   (b) There are twice as many nickels as quarters.
   (c) There are 45 coins in all.

4. A quarter is worth 25¢, a dime 10¢, and a nickel 5¢.

5. Let
$$q = \text{the number of quarters Arnold has}$$
$$d = \text{the number of dimes he has}$$
   Then
$$2q = \text{the number of nickels}$$

6. Notice that at Step 5 we built in the data stated in 3(b). The total value is
$$25q + 10d + 5(2q) = 35q + 10d$$
   and the total number of coins is
$$q + d + 2q = 3q + d$$

7. From 3(a) we get
$$35q + 10d = 500 \qquad (1)$$
   and from 3(c) we get
$$3q + d = 45 \qquad (2)$$

8. Dividing equation (1) by 5 we get
$$7q + 2d = 100$$
   and multiplying equation (2) by 2 we get
$$6q + 2d = 90$$
   Then subtracting these last two equations gives
$$q = 10 = \text{the number of quarters}$$
   Then
$$2q = 20 = \text{the number of nickels}$$
   and from equation (2)
$$d = 45 - 3q = 45 - 30 = 15 = \text{the number of dimes}$$

Number and Money Problems

9. We check 3(a);

$$10 \times 25 + 15 \times 10 + 20 \times 5 = 250 + 150 + 100 = 500$$

then 3(b),

$$20 = 2 \times 10$$

and finally 3(c),

$$10 + 15 + 20 = 45$$

EXERCISE 3   Amos has $7.35 in half dollars, quarters and dimes. The number of quarters is one less than twice the number of dimes, and he has twenty-five coins in all. How many of each kind of coin does he have?

ADDITIONAL EXERCISES

4. Tom has five dollars in nickels and dimes. How many of each does he have if there are twice as many dimes as nickels?

5. If in Exercise 4, Tom had had twice as many nickels as dimes, how many of each would he have?

6. In a two-digit number the tens' digit is three more than the units' digit. The number is three less than seven times the number. Find the number.

7. In a two-digit number the square of the units' digit is two more than twice the tens' digit, and twice the units' digit is one more than the tens' digit. Find the number.

8. The difference of the squares of two consecutive integers (whole numbers) is 61. What are the numbers? Question: What does "consecutive" mean?

9. Find two positive integers whose product is 84 and the difference of whose squares is 95.

10. The difference of the reciprocals of two consecutive positive integers is $\frac{1}{30}$. Find the numbers. Question: What does "reciprocal" mean?

6
GEOMETRY PROBLEMS

Here for the first time Step 4 begins to be quite important because many geometry problems stated in elementary algebra assume that certain theorems from geometry are familiar. Also, for the first time, the drawing of figures in Step 6 and Step 7 becomes very important. Remember, whenever possible, *always draw a figure* illustrating the problem.

EXAMPLE 1   Way back in the year 2070 a moving conveyor belt was run around the earth at the equator. Then, because the polar ice caps were melting and raising the sea level it became necessary in 2170 to raise the belt so that it ran 100 feet above its former level. To do this required lengthening the belt, so it was cut and a section of the proper length added. How long was the section? (Assume the earth a perfect sphere.)

Solution
1. READ THE PROBLEM.
2. The length of the added section is asked for.
3. A belt around the earth is lengthened so that it can be raised 100 feet.
4. The circumference of a circle is 2π times its radius:
$$C = 2\pi r$$
5. Let
$$x = \text{length of the added section}$$

Geometry Problems

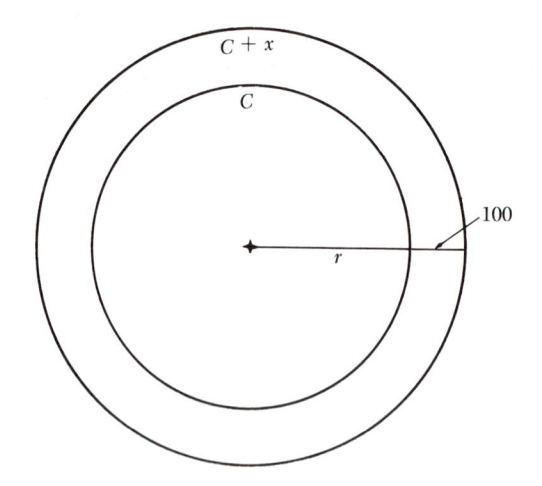

[Figure: Two concentric circles. Inner circle labeled with circumference C and radius r. Outer circle labeled with circumference $C+x$, with the outer radius exceeding the inner by 100.]

6. From the figure it is clear that lengthening the radius by 100 ft lengthens the belt by x ft.

7.
$$C = 2\pi r \tag{1}$$
and
$$C + x = 2\pi(r + 100) = 2\pi r + 200\pi \tag{2}$$

8. Subtracting equation (1) from (2) gives

$$x = 200\pi$$

9. Equation 2 verifies the solution.

Notice that the solution is independent of the size of the original circle. What we have done really is to show that if the radius of *any* circle is increased by 100 feet, then the circumference increases by 200π feet.

EXERCISE 1   Prove that if the radius of a circle is increased by f feet then the circumference is increased by $2\pi f$ feet.

EXAMPLE 2   An architect is designing triangular windows for a new church. He has a basic pattern from which he works, and for this pattern the base is two feet more than the altitude. If he increases

Geometry Problems

the altitude by three feet the area increases by nine square feet. Find the original altitude.

Solution

1. READ THE PROBLEM.

2. The altitude of the basic pattern triangle is sought.

3. (a) The base is two feet more than the original altitude.
   (b) An increase of three feet in the altitude causes an increase of nine square feet in area.

4. The area of a triangle is one-half of the base times the altitude:

$$A = \frac{1}{2}bh$$

5. Let
   $h =$ the original altitude in feet, then
   $h + 2 =$ the length of the base in feet, and
   $h + 3 =$ the new altitude in feet.

6.
$$\frac{1}{2}(h+2)h = \text{original area in square feet}$$
$$\frac{1}{2}(h+2)(h+3) = \text{new area in square feet}$$

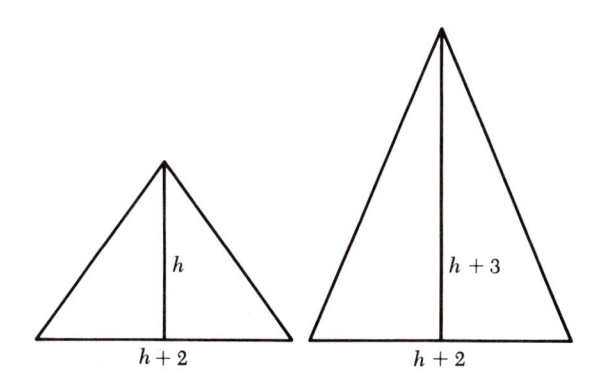

7.
$$\frac{1}{2}(h+2)(h+3) - \frac{1}{2}(h+2)h = 9$$

36   Geometry Problems

8.
$$(h^2 + 5h + 6) - (h^2 + 2h) = 18$$
$$3h = 12$$
$$h = 4 = \text{original altitude in feet}$$

9. From Step 5 the length of the base is 6 ft. The original is then ½ · 4 · 6 = 12. The new altitude is 4 + 3 = 7, so the new area is ½ · 7 · 6 = 21.

$$21 - 12 = 9$$

EXERCISE 2 A rectangle has one side three feet longer than the other. If the long side is lengthened by 2 feet the area is increased by 6 feet. Find the length of the sides of the original rectangle.

EXAMPLE 3 A fisherman wants to know how deep the water is. He pulls his boat up beside a bulrush which is sticking up one foot above the water. When he pulls it sideways so that the top of the rush is level with the water it is at a point five feet from where it first was. Find the depth.

Solution

1. READ THE PROBLEM.

2. The depth of the water is wanted.

3. (a) The rush sticks up one foot above the water level.
   (b) When pulled to one side 5 feet the top of the rush is level with the water.

4. The Theorem of Pythagoras says that for a right triangle the square of the hypotenuse is the sum of the squares of the other two sides (legs).

5. Let
$$x = \text{depth of water in feet, then}$$
$$x + 1 = \text{length of rush}$$

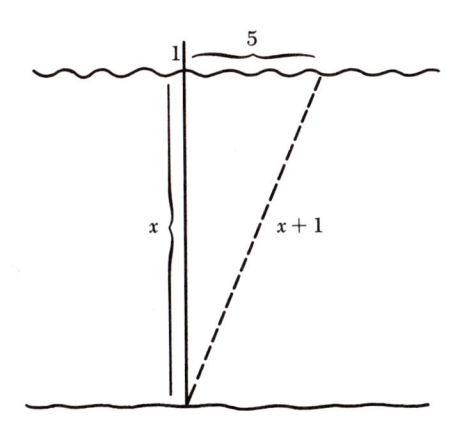

6. From the figure we see that the water level, the old position, and the new position of the bulrush form a right triangle.

7. From Step 4 (Pythagoras's Theorem) we get
$$(x + 1)^2 = x^2 + 5^2$$

8.
$$x^2 + 2x + 1 = x^2 + 25$$
$$2x = 24$$
$$x = 12 = \text{depth of water in feet}$$

9. The length of the rush is 13 feet:
$$13^2 = 169, \quad 12^2 = 144, \quad 5^2 = 25, \quad 169 = 144 + 25$$

EXERCISE 3 A flagpole has a rope attached to its tip which is four feet longer than the pole itself. The rope stretches out to a point on the level ground 24 feet from the base of the pole. Find the height of the pole and the length of the rope.

EXAMPLE 4 A right circular cone of altitude h and having a base of radius r is inscribed in a sphere of radius 6. The volume of the cone is $h/54$ times the volume of the sphere. Find h and r.

38   Geometry Problems

Solution

1. READ THE PROBLEM.

2. The altitude and radius of the base of the cone are asked for.

3. (a) The cone is inscribed in a sphere of radius 6.
   (b) The volume of the cone is $h/54$ times the volume of the sphere.

4. The volume of a sphere is $4\pi R^3/3$ where R is the radius of the sphere.
   The volume of a cone is ⅓ Bh where h is the altitude and B is the area of the base

$$\frac{1}{3} Bh = \frac{1}{3} \pi r^2 h$$

where r is the radius of the base.

5. These were given in the problem.

6. The volume of the sphere is

$$\frac{4\pi 6^3}{3} = 288\pi$$

The volume of the cone is

$$\frac{1}{3} \pi r^2 h$$

7. From 3(b) we get

$$\frac{1}{3} \pi r^2 h = \frac{h}{54} \cdot 288\pi$$

8. Dividing by πh and multiplying by 3 gives

$$r^2 = 16$$

or

$$r = \pm 4$$

and we use

$$r = 4 = \text{radius of the base of the cone}$$

We have not yet drawn a figure, because we have not needed it. But now in order to find h we do:

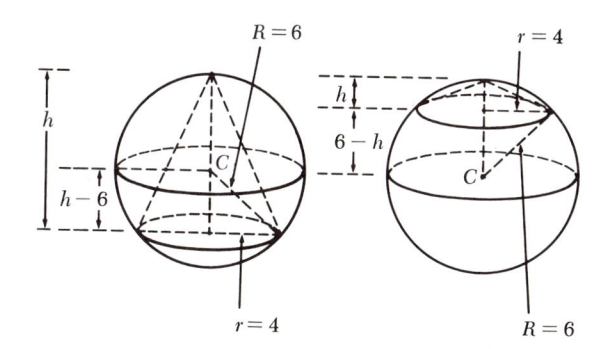

We recognize two possibilities. If the cone is placed with its vertex at the top of the sphere, there are two altitudes, h, which permit it to have a base of radius 4. In one case $h > 6$ and the base lies below the equatorial plane of the sphere; in the other $h < 6$ and it lies above. Then in either case

$$(h-6)^2 + 4^2 = 6^2$$

(why?)
so that
$$(h-6)^2 = 36 - 16 = 20 = 4 \times 5$$
$$h - 6 = \pm 2\sqrt{5}$$
$$h = 6 \pm 2\sqrt{5} = 2(3 \pm \sqrt{5})$$

9. We leave the solution for the student to check.

EXERCISE 4 A right circular cone has altitude 80 cm and a base of radius 60 cm. A right circular cylinder is inscribed so that a base of the cylinder sits on the base of the cone, and both the cylinder and the cone have the same axis. If the total area of the cylinder (i.e., side and bases) is equal to 0.22 times the total area of the cone, find the dimensions of the cylinder.

ADDITIONAL EXERCISES

5. Three circles are drawn with centers at points A, B, and C. The circles with centers B and C are inside the one with center at A, and are internally tangent to it, and they are externally tangent to each other. The distance AB is seven inches, AC is eight inches, and BC is 9 inches. Find the radii of the circles.

6. Two squares of different sizes have the sum of their perimeters 56 feet. When they are placed together so that a side of the smaller square lies along one side of the larger, they form a figure with perimeter 45 feet. Find the sizes of the squares.

7. A rectangular picture frame is ¾ inch wide along the sides and one inch wide at the top and bottom. It holds a picture of area 24 square inches, and the area of the frame itself is 20 square inches. Find the outside dimensions of the frame.

8. A landscape architect wants to install a reflecting pool with area 384 square feet in a space 20 feet wide and 40 feet long. If he leaves a walkway around the pool of the same width all around, what dimensions does the pool have?

9. A rectangular cardboard is twice as long as it is wide. A square 1½ inches on a side is cut from each corner and the ends bent up to form a box with a volume of 21 cubic inches. Find the dimensions of the cardboard.

10. The rear wheels of a dragster are one foot more around than the front wheels. In a one-mile drag strip the front wheels rotate 40 more times than the rear ones. Find the circumferences of both sets of wheels.

7
WORK PROBLEMS

EXAMPLE 1   Arnold, the house painter, can paint a certain wall in three hours, and his young assistant, Bert, can paint the same wall in four hours. How long will it take Arnold and Bert working together to paint the wall?

Before beginning the solution of this problem some explanatory comments are in order. It is assumed that the work is done at a steady rate, that is, in one hour Arnold does ⅓ of the job, and Bert does ¼. In two hours Arnold does ⅔ and Bert does ½, etc.

Note that the fractions ¼, ⅓, ½, and ⅔ in the previous paragraph represent the *fractional part* of the *whole job* which is done by Arnold or Bert in the times mentioned. When all of the job is done, the fraction is 1. This is similar to percentage: ⅓ of the job is done when 33⅓% is done, and ½ when 50% is done. The whole job is done when 1 or 100% of the job is done. Thus these fractional parts are = percentage/100.

Solution

1. READ THE PROBLEM.

2. The time for Arnold and Bert working together to paint the wall is asked for.

3. Arnold can paint the wall in 3 hours; Bert can paint the wall in 4 hours.

4. This was covered in the comments.

5. Let
$x =$ the number of hours required

6. In one hour

Arnold does $\frac{1}{3}$ of the work

Bert does $\frac{1}{4}$ of the work

so A and B together do

$$\frac{1}{3} + \frac{1}{4} = \frac{7}{12}$$

of the work. But if together they do the work in x hours then in one hour they do $1/x$ of the work.

7. Then

$$\frac{1}{x} = \frac{7}{12}$$

8.
$$x = \frac{12}{7} = 1\frac{5}{7} \text{ hr}$$

9.
$$\frac{12}{7}\left(\frac{1}{3} + \frac{1}{4}\right) = \frac{12}{7}\left(\frac{7}{12}\right) = 1$$

EXERCISE 1   Arnold and his other assistant Charlie are painting another wall. Arnold could paint it alone in 10 hours, and together Arnold and Charlie take 6 hours to complete the work. How long would it take Charlie by himself to paint the wall?

EXAMPLE 2   The tank of a liquid fuel rocket can be filled from one pipe connection in 6 hours or from another in 8 hours. If both pipes work together for 2 hours, how long will it take the slower one to then finish filling the tank?

Solution

1. READ THE PROBLEM.
2. The time required for the slower pipe to finish is required.
3. (a) One pipe could fill in 6 hours.
   (b) Another could fill in 8 hours.
   (c) They both work together for 2 hours.
4. See the comments between the statement and the solution of Example 1 of this chapter.

Work Problems 43

5. Let
    $x =$ the number of hours for the slower pipe to finish filling the tank

6. From 3(a), the faster pipe fills ⅙ of the tank in one hour. From 3(b), the slower one fills ⅛ of the tank in one hour. Together they fill

$$\frac{1}{8} + \frac{1}{6} = \frac{7}{24} \text{ of the tank in one hour}$$

7.
$$2\left(\frac{7}{24}\right) + x\left(\frac{1}{8}\right) = 1 \quad \text{(why 1?)}$$

8.
$$\frac{14}{24} + \frac{x}{8} = 1$$
$$14 + 3x = 24$$
$$3x = 10$$
$$x = 3\frac{1}{3}\text{hr} = 3 \text{ hr } 20 \text{ min}$$

9. Carry out the checking.

EXERCISE 2   A tank can be filled by one pipe in 4 hours, by another in 6, and can be emptied by a third in 5 hours. How long will it take to fill the tank if all pipes are open?

EXAMPLE 3   Two pipes working together can fill a tank in 3 hours. But if the faster one is filling and the slower one draining, it takes 12 hours. How long does it take each pipe alone to fill the tank? (Assume the pipe drains at the same rate it would fill.)

Solution
1. READ THE PROBLEM.
2. The filling time for each pipe is asked for.
3. (a) Working together they take 3 hours to fill.
   (b) With the slower one draining it takes 12 hours to fill.

44   Work Problems

4. See the comments between the statement and the solution of Example 1 of this chapter.

5. Let
   x = the number of hours for the faster pipe to fill, and
   y = the number of hours for the slower one to fill.

6. $\dfrac{1}{x}$ = the fractional part of the tank filled by faster pipe in one hour,

   $\dfrac{1}{y}$ = the fractional part of the tank filled by the slower pipe in one hour.

7.
$$\frac{1}{x} + \frac{1}{y} = \frac{1}{3}$$
$$\frac{1}{x} - \frac{1}{y} = \frac{1}{12}$$

8. If we clear the equation of fractions we have problems: the xy term. If we treat $1/x$ and $1/y$ as our unknowns, the solution is easy: adding the equations we get

$$\frac{2}{x} = \frac{1}{3} + \frac{1}{12} = \frac{5}{12}$$
$$\frac{1}{x} = \frac{5}{24}$$
$$x = \frac{24}{5} = 4\frac{4}{5} \text{ hr}$$
$$= 4 \text{ hr } 48 \text{ min}$$

Subtracting we get

$$\frac{2}{y} = \frac{1}{3} - \frac{1}{12} = \frac{3}{12} = \frac{1}{4}$$
$$\frac{1}{y} = \frac{1}{8}$$
$$y = 8 \text{ hr}$$

EXERCISE 3   Three pipes lead into a tank. The first and second working together fill it in 3 hours. If the first and third are filling, and the second draining, it takes 6 hours to fill. If the first and third fill at the same rate, how long would it take each working alone to fill the tank?

ADDITIONAL EXERCISES

4. Arnold and a third assistant, Don, working together can paint a wall in 2 hours and 24 minutes, but working alone it would take Don two hours more than it would Arnold. How long would each take to paint the wall alone?

5. The earth moves all the way around the sun in about 365 days, while Venus takes about 225 days. How long is it from the time these planets are together on a radius from the sun until they are next together?

8
MIXTURE AND VALUE PROBLEMS

Most mixing problems can be set up by means of "balance" equations expressing the fact that the whole is the sum of each of its parts for each component of the mixture. The same sort of reasoning applies to the "value balance" of the type used in the oil problem we solved in Chapter 1, the first problem worked in this booklet.

EXAMPLE 1   A chemical company has 3600 tons of a solution of 40 percent saccharin. How much of a solution of 75 percent saccharin must be added to bring it to 50 percent saccharin?

Solution

1. READ THE PROBLEM.

2. How much 75 percent saccharin should be added to get a 50 percent solution?

3. (a) 3600 tons 40 percent saccharin.
   (b) Add solution of 75 percent saccharin.
   (c) Mixture has 50 percent saccharin.

4. Nothing here.

5. Let
    x = number of tons of 75 percent saccharin added.

6. Heading toward a saccharin balance:

    $3600(.40)$ = number of tons of saccharin originally
    $x(.75)$ = number of tons of saccharin added
    $(3600 + x)(.50)$ = number of tons of saccharin in the final mixture

7.
$$3600(.40) + x(.75) = (3600 + x)(.5)$$

48 Mixture and Value Problems

8.
$$1440 + .75x = 1800 + .5x$$
$$.25x = 360$$
$$x = 1440 \text{ tons of } 75\% \text{ saccharin added}$$

9.
$$3600(.40) + 1440(.75) = 1440 + 1080 = 2520$$
$$5040(.50) = 2520$$

EXERCISE 1 How many milligrams of uranium 72% pure and of uranium 84.8% pure must be mixed to give 8 milligrams of uranium 80% pure?

EXAMPLE 2 A biologist has 3 grams of some mouse food containing .006% strontium-90 to be used as a radioactive trace in a study of bone cancer in mice. How many grams of mouse food containing .011% strontium-90 must he add to get a food containing .008% strontium-90?

Solution

1. READ THE PROBLEM.

2. The number of grams of food with .011% strontium-90 is sought.

3. (a) There are 3 grams of .006% strontium-90.
   (b) Some .011% strontium-90 is added.
   (c) The mixture should be .008% strontium-90.

4. Nothing here.

5. Let
   $x =$ the number of grams of .011% strontium-90 added

6. From 3(a):

   $3(.00006) =$ the number of grams of pure strontium-90 in the 3g of food

   From 3(b) and 5:

   $x(.00011) =$ the number of grams of pure strontium-90 added

From 3(c) and 5:

$(3 + x)(.00008)$ = the total number of grams of pure strontium-90

7. Since the whole is equal to the sum of its parts:

$$3(.00006) + x(.00011) = (3 + x)(.00008)$$

8.
$$.00018 + .00011x = .00024 + .00008x$$
$$.00003x = .00006$$
$$x = 2$$

9.
$$.00018 + .00022 = 5(.00008)$$
$$.00040 = .00040$$

EXERCISE 2 In a tropical fish store, a tank contains 65 lb of sea water containing 7% salt. (a) How much pure water should be added to get 5% salt? (b) How much water should be evaporated off to get 9% salt?

EXAMPLE 3 A candy store has two kinds of candy. One kind sells for $1.00 a pound and another for $1.75 a pound. How should it be mixed so as to sell for $1.30 a pound?

Solution
1. READ THE PROBLEM.

2. No specific number is asked for, but how to mix two kinds of candy. To answer, we can find how much of each kind of candy goes into a pound of the mixture.

3. (a) One kind of candy is worth $1.00 a pound.
   (b) The other is worth $1.75 a pound.
   (c) The mixture is worth $1.30 a pound.

4. Nothing here.

50   *Mixture and Value Problems*

5. Let

   $x =$ the number (fractional, of course) of pounds of $1.00 candy in each pound of mix

   Then

   $1 - x =$ the number of pounds of $1.75 candy in the mix

6. Aiming for a value balance

   $(1.00)x = x =$ the value of $1.00 candy in each pound of mix

   Then

   $(1.75)(1 - x) =$ the value of $1.75 candy in each pound of the mix

7. Then since the whole is the sum of its parts,

   $$x + 1.75(1 - x) = 1.30$$

8.

$$100x + 175(1 - x) = 130$$
$$175 - 75x = 130$$
$$75x = 45$$
$$x = \frac{45}{75} = \frac{3}{5}$$
$$1 - x = \frac{2}{5}$$

   Each pound of the mix should contain

   $\frac{3}{5}$ pound of $1.00 candy

   $\frac{2}{5}$ pound of $1.75 candy

9.

$$\left(\frac{3}{5}\right)(1.00) + \left(\frac{2}{5}\right)(1.75) = \frac{3}{5} + \frac{3.50}{5} = \frac{6.50}{5} = \frac{13.0}{10} = \frac{130}{100} = 1.30$$

EXERCISE 3   Betty likes chocolates that cost $2.00 a pound and her little brother Arnold likes hard candy that costs 75 cents a pound. They have a dollar and a half. If they buy a pound of a mixture of the two kinds of candy, how much of each kind will they get?

Mixture and Value Problems 51

EXAMPLE 4 Ten pounds of an alloy contain 5% silver and 40% platinum. How much pure silver and how much pure platinum must be added to make an alloy containing 20% silver and 40% platinum?

Solution

1. READ THE PROBLEM.
2. The number of pounds of silver and platinum added is sought.
3. (a) The original alloy contains 5% silver.
   (b) The original alloy contains 40% platinum.
   (c) The new alloy contains 20% silver.
   (d) The new alloy contains 40% platinum.
4. Nothing here.
5. Let

   $x =$ the number of pounds of silver added
   $y =$ the number of pounds of platinum added

6. We are aiming toward a silver balance and a platinum balance. First we note that since we have added x pounds of silver and y pounds of platinum, the total weight of the new alloy in pounds is

   $$10 + x + y$$

   (a) The amount (pounds) of silver in the old alloy was

   $$(.05)(10) = .5$$

   (b) The amount (pounds) of platinum in the old alloy was

   $$(.40)(10) = 4$$

   (c) The amount (pounds) of silver in the new alloy is

   $$(.20)(10 + x + y)$$

   (d) The amount (pounds) of platinum in the new alloy is

   $$(.40)(10 + x + y)$$

7. The silver balance equation is

   $$.5 + x = .2(10 + x + y) \qquad (1)$$

   and the platinum balance equation is

   $$4 + y = .4(10 + x + y) \qquad (2)$$

52 Mixture and Value Problems

8. Then from (1)
$$.8x - .2y = 2 - .5 = 1.5$$
or
$$8x - 2y = 15 \qquad (3)$$
and from (2)
$$.4x - .6y = 0$$
or
$$4x - 6y = 0 \qquad (4)$$
Then multiplying equation (4) by (2) and subtracting from (3), we get
$$10y = 15$$
$$y = 1.5 = \text{pounds of platinum added}$$
From this we get
$$x = 2.25 = \text{pounds of silver added}$$

9. We leave it to the reader to check this solution.

EXERCISE 4 A stockroom attendant in a laboratory has weak hydrochloric acid at 5%, stronger at 25%, and still stronger at 60%. He makes a mixture of all three yielding 230 grams of 30% hydrochloric acid. He used three times as much of the 25% acid as of the 60% acid. How much of each was used?

EXAMPLE 5 A biologist wants to determine the volume of blood in a small animal. He has a solution of which 4% is a certain biologically inert chemical. He injects 2.5 cubic centimeters of this solution into the animal's bloodstream, and waits until it is well mixed. He then withdraws a small sample. Analysis shows that 0.2% of the sample is the chemical. How much blood does the animal have?

Solution
1. READ THE PROBLEM.
2. The volume of blood in the animal is wanted.
3. (a) 2.5 cc of fluid is injected.
   (b) 4% of the injection is the chemical.
   (c) 0.2% of the blood sample is the chemical.

Mixture and Value Problems 53

4. Nothing here—except that it is to be understood that the solution is well mixed in the blood so that the 0.2% holds for all the blood.

5. Let
$$V = \text{the volume (number of cc) of blood}$$

6. Then
$$V + \frac{5}{2} = \text{the volume of the blood and the solution after the injection}$$

Now the amount of the chemical in the blood after injection is equal to the amount injected. The amount (number of cc) of the chemical injected is

$$(.04)\left(2\frac{1}{2}\right) = \frac{1}{25} \cdot \frac{5}{2} = \frac{1}{10} = 0.1$$

and the amount (number of cc) in the bloodstream is

$$(.002)\left(V + \frac{5}{2}\right) = .002V + .005$$

7. Then
$$0.1 = .002V + .005$$

8.
$$.002V = 0.1 - .005 = .095$$
$$.002V = .095$$
$$V = \frac{.095}{.002} = \frac{95}{2} = 47.5$$
$$= \text{volume (in cc) of blood}$$

9. We leave this for the student.

ADDITIONAL EXERCISES

5. Betty's candy store has one kind of candy for 75¢ a pound and another for $1.00 a pound. How much of the $1.00 candy should be added to five pounds of the 75¢ candy so that the mixture is worth 90¢ a pound?

6. Albert goes into Betty's store and wants three pounds of candy for $2.60. How much 75¢ and how much $1.00 candy should he get?

Mixture and Value Problems

7. Bill drives into Andy's garage. Andy tests Bill's radiator and finds it holds 16 quarts and is full of 30% ethylene glycol. How much should he drain out and replace with pure ethylene glycol to have 40% ethylene glycol in the radiator?

8. A garbage company has three helpers for each two drivers. The drivers get $5.00 an hour, and the helpers $2.00 an hour each. With an 8-hour day and a 5-day week the payroll is $3840. How many drivers and helpers are there?

9
INTEREST PROBLEMS

If Arnold loans money to Bert, or deposits it in Juan's savings bank, then Bert or Juan pays Arnold for the use of his money. When Bert repays the loan, or Arnold withdraws his money from Juan's bank, Arnold gets back his loan or deposit, plus some extra. The amount of the loan is called the *principal* and the extra amount is called the *interest*.

In the simplest situation, where the loan is for one year, the interest is some previously agreed percentage (called the *rate* of interest) of the principal. Thus if Arnold had deposited $125 in Juan's bank at 5% per year interest, he would have

$$5\% \text{ of } 125 = .05 \times 125 = 6.25$$

dollars in interest or a total of

$$125 + 6.25 = 131.25$$

dollars at the end of the year.

We can write a formula to describe this. In general if I represents the amount of the interest, P the amount of the principal, and r the rate, expressed in fractional parts rather than percentage, i.e.,

$$r = (\text{percentage rate}) \div 100$$

then,

$$I = Pr$$

and, at the end of the year the total money, including principal and interest, is

$$P + I = P + Pr = P(1 + r)$$

If, at the end of the year, this total amount is re-invested at the same rate, then the new principal is $P(1 + r)$. The interest on the second year is then

$$P(1 + r)r$$

and the total amount at the end of the second year is

$$P(1 + r) + P(1 + r)r = P(1 + r)(1 + r) = P(1 + r)^2$$

55

56   Interest Problems

If this is done every year for n years, the interest on the last year is
$$P(1+r)^{n-1}r$$
and the total amount at the end of the last year is
$$P(1+r)^n$$
This procedure is called *compounding* the interest, and the last formula for the total amount is called the *compound interest formula*.

EXAMPLE 1 Arnold loans some money to Bert at 4% per year, and deposits three times that much in Juan's bank at 5%. At the end of the year Arnold's total interest is $1330. How much did he lend Bert, and how much did he deposit in Juan's bank?

Solution
1. READ THE PROBLEM.
2. The amount of each investment is asked for.
3. (a) The loan is at 4%.
   (b) The savings account is at 5%.
   (c) The deposit is three times the loan.
   (d) The total interest is $1330.
4. Nothing here.
5. Let
    x = amount (number of dollars) Arnold loaned Bert
   Then
    $3x$ = amount (in dollars) of deposit
6. The interest on the loan is
    $$(.04)x = .04x \text{ dollars}$$
   and the interest on the savings is
    $$(.05)\,3x = .15x \text{ dollars}$$
7. Then since the total interest is the sum of the separate interests
    $$.04x + .15x = 1330$$

8.
$$.19x = 1330$$
$$19x = 133000$$
$$x = 133000 \div 19$$
$$x = 7000 \text{ dollars}$$
$$= \text{the loan to Bert}$$
$$3x = 21{,}000 = \text{the savings account}$$

9.
$$\begin{array}{r} 7{,}000 \times .04 = 280 \\ 21{,}000 \times .05 = 1050 \\ \hline \text{Total} = 1330 \end{array}$$

EXERCISE 1 A man has $10,000 invested, part at 4% and part at 6%. If his yearly income is $520, how much is invested at each rate?

EXAMPLE 2 Hank invests $100 at compound interest. In two years he has $112.36. What is the yearly rate of interest?

Solution
1. READ THE PROBLEM.
2. The rate of interest is asked for.
3. (a) There was originally $100.
    (b) It is invested at compound interest.
    (c) In two years there is $112.36.
4. Nothing here beyond the explanation at the beginning of this chapter.
5. Let
$$r = \text{the rate of interest}$$
6. By the compound interest formula the amount in two years is
$$100(1 + r)^2$$
7.
$$100(1 + r)^2 = 112.36$$

58   Interest Problems

8.
$$(1 + r)^2 = 1.1236$$

Taking square root:
$$1 + r = 1.06$$
$$r = .06$$

The rate is then 6%.

9.
$$100(1.06)^2 = 100(1.1236) = 112.36$$

EXERCISE 2 George invests $100 at compound interest for two years. He has $110.25 at the end of the two years. What is the yearly rate of interest?

EXAMPLE 3 Tom has two investments totaling $26,000, part at 5% and part at 4.5%. The part at 5% pays $350 per year more than that at 4.5%. Find the principal of each.

Solution
1. READ THE PROBLEM.
2. The principals of two investments are asked for.
3. (a) The sum of the principals is $26,000.
   (b) One investment is at 5% interest.
   (c) The other is at 4.5% interest.
   (d) The 5% investment earns 350 per year more than the 4.5% one.
4. Nothing here.
5. Let
$$x = \text{the principal invested at } 4.5\%$$
   Then
$$26{,}000 - x = \text{the principal invested at } 5\%$$
6. Then
$$.045x = \text{interest from one investment}$$
$$.05(26{,}000 - x) = \text{interest from the other}$$

7. From Step 3(d) and Step 6:
$$.05(26{,}000 - x) - .045x = 350$$

8.
$$1300 - .095x = 350$$
$$.095x = 950$$
$$x = 10{,}000 = \text{amount at } 4.5\%$$
$$26{,}000 - x = 16{,}000 = \text{amount at } 5\%$$

9. We leave this to the student.

EXERCISE 3 Gerald has two investments, one twice as large as the other. At the end of the year they have earned $800. What were the two investments if the interest was 6% on the larger investment and 4% on the smaller?

EXAMPLE 4 Tony gets $48.00 a year from an investment. If he had another $300 invested and if the rate were 0.5% less, he would get $52.50. How much does he have invested and what is the interest rate?

Solution

1. READ THE PROBLEM.
2. Both the principal and the rate are asked for.
3. (a) The interest is $48.00.
   (b) If (i) the principal were $300 more and
       (ii) the rate were 0.5% less, then
       the interest would be $52.50.
4. Nothing here.
5. Let
$$r = \text{the rate of interest}$$
$$P = \text{the principal (in dollars)}$$
6. Then
$$P + 300 = \text{changed principal}$$
$$(r - .005) = \text{changed rate}$$
7.
$$Pr = 48$$
$$(P + 300)(r - .005) = 52.50$$

Interest Problems

8. Then,
$$Pr + 300r - .005P - 1.5 = 52.50$$

Subtracting the first equation of Step 7 we get

$$300r - .005P - 1.5 = 4.5$$
$$300r = .005P + 6 = .005\,(P + 1200)$$
$$300r = .005P + 6 = .005\,(P + 1200)$$
$$300r = \frac{1}{200}\,(P + 1200)$$
$$r = \frac{1}{60{,}000}\,(P + 1200)$$

We substitute this back into the first equation of Step 7:

$$\frac{P(P + 1200)}{60{,}000} = 48$$
$$P^2 + 1200P - 2{,}880{,}000 = 0$$
$$(P + 2400)(P - 1200) = 0$$
$$P = \$1200 = \text{principal}$$

Again from the first equation of Step 7

$$1200r = 48$$
$$r = \frac{48}{1200} = .04$$

So the rate is 4%.

9.
$$.04 \times 1200 = 48$$
$$.035 \times 1500 = 52.50$$

10
RATE PROBLEMS

You probably have a pretty good idea what we mean by "speed," or "rate of speed," or "velocity"—these all meaning essentially the same thing. But for the sake of clarity we now review the main ideas.

Suppose we are observing an object moving in a straight line, for example, a car moving along a road. If we watch the motion of the car for a certain length of time, T (we have a stop-watch so we can measure time), and during this time the car has moved a distance, D, down the road. Then we say that the *average speed*, which we will call R (for rate), for that part of the motion, is given by

$$R = \frac{D}{T} \tag{1}$$

That is, we *define* the average speed by this formula. From (1) we get the equivalent equations

$$D = RT \tag{2}$$

and

$$T = \frac{D}{R} \tag{3}$$

It is clear that for a typical trip in an automobile, the average speed, for different parts of the trip, would vary widely. However, after we get the car moving, we can maintain approximately a steady or constant speed on open stretches of road. To be precise, we define *constant speed* as follows: If an object is moving so that the *average speed* is the same for *all parts* of the motion, we say the object is moving with *constant speed*. Of course, in practice, we can usually expect only to approximate constant speed.

It follows from the definition of speed or average speed in equation (1) that the units of speed are in "distance" divided by "time": miles per hour, feet per second, meters per century, etc. And it also follows that for any of the formulas (1), (2) and (3) to make sense the units of time and distance must be the same in R as in D and T. Thus if R is

62   Rate Problems

miles per hour and D is in feet, then before they can be used in either of these formulas they must be converted to the same unit of distance. The same remark applies of course to the units of time.

EXAMPLE 1   A jet and a propeller plane are going in the same direction, with the jet going twice as fast as the propeller plane. Three hours after they leave, the jet is 861 miles ahead. How fast is each plane going?

Solution

1. READ THE PROBLEM.
2. The speeds of the two planes are sought.
3. (a) They go in the same direction.
   (b) Though it is not stated directly it is implied that they leave at the same time.
   (c) One is twice as fast as the other.
   (d) They travel three hours.
   (e) Then they are 861 miles apart.
4. Nothing here.
5. Let

   $x =$ speed in miles per hour of the propeller plane

   $2x =$ speed in miles per hour of the jet

6. The distance the jet travels is

   $3(2x) = 6x$ miles

   The distance the propeller plane travels is

   $3(x) = 3x$ miles

7. The difference of their distances is 861 miles:

   $6x - 3x = 861$

8. We solve for x:

   $3x = 861$

   $x = \dfrac{861}{3} = 287$ miles per hour,
   the speed of the propeller plane

   $2x = 574$ miles per hour,
   the speed of the jet

Rate Problems 63

9. Check:

$$3 \times 574 = 1722$$
$$3 \times 287 = 861$$
$$\text{Subtract: } 861$$

EXERCISE 1 A hiker and a bicyclist are going along the same trail, with the bicyclist going three times as fast as the hiker. They start at 8:00 A.M. and at 12:00 noon the cyclist is 32 miles ahead of the hiker. How fast is each going?

EXAMPLE 2 On a 45 kilometer trip in and around Paris a Renault went at 50 kilometers per hour (km.p.hr.) outside the city and 30 km.p.h. inside. If the trip took one hour, how many kilometers were traveled inside the city?

Solution
1. READ THE PROBLEM.
2. How many kilometers were traveled within the city?
3. (a) Total distance: 45 km
   (b) Total time: 1 hr
   (c) Rate in city: 30 km.p.hr.
   (d) Rate outside city: 50 km.p.hr.
4. Nothing here.
5. Let

    $x =$ number of kilometers traveled inside the city

   Then

   $45 - x =$ number of kilometers traveled outside the city
6. By formula (3)

$$T = \frac{D}{R}$$

$$\frac{x}{30} = \text{time (hours) inside the city}$$

$$\frac{45 - x}{50} = \text{time (hours) outside the city}$$

64   Rate Problems

7. By Step 6 and Step 3(b):
$$\frac{x}{30} + \frac{45-x}{50} = 1$$

8.
$$50x + 1350 - 30x = 1500$$
$$5x + 135 - 3x = 150$$
$$2x = 15$$
$$x = \frac{15}{2} = 7\frac{1}{2}$$
$$45 - x = \frac{75}{2} = 37\frac{1}{2}$$

9.
$$\text{Time inside city} = \frac{7\ 1/2}{30} = \frac{15/2}{30} = \frac{15}{60} = \frac{1}{4}$$
$$\text{Time outside city} = \frac{75/2}{50} = \frac{75}{100} = \frac{3}{4}$$
$$\text{Total time} = \frac{1}{4} + \frac{3}{4} = 1$$

Redo this problem by letting, in Step 5,
$$t = \text{time (hours) inside the city}$$
$$(1-t) = \text{time outside the city}$$

EXERCISE 2   How soon after 12 o'clock will the hands of a clock point in the same direction again?

EXAMPLE 3   Anthony can run around a track in 40 seconds, and his little brother Bill can run around it in 50 seconds. If Anthony gives Bill a 3-second head start how long does it take Anthony to catch up with Bill?

Solution
1. READ THE PROBLEM.
2. The time for Anthony to catch Bill is asked for.
3. (a) Anthony can run around the track in 40 sec.
   (b) Bill can run around the track in 50 sec.

Rate Problems 65

(c) Bill has a 3-sec head start.
(d) When Anthony catches Bill they will have run the same distance.

4. Nothing here.
5. Let

$D =$ the number of yards around the track, and
$T =$ the number of sec for Anthony to catch Bill.

Then

$T + 3 =$ the number of sec Bill runs before Anthony catches up.

6. From 3(a):

$$\frac{D}{40} = \text{Anthony's speed (yards per sec),}$$

and from 3(b)

$$\frac{D}{50} = \text{Bill's speed (yards per sec)}$$

7.

$$\frac{D}{40} \times T = \text{the distance Anthony runs to catch Bill}$$

$$\frac{D}{50}(T + 3) = \text{the distance Bill runs before Anthony catches him}$$

From 3(d):

$$\frac{DT}{40} = \frac{D(T + 3)}{50}$$

8. We multiply by 200 and divide by D:

$$5T = 4(T + 3)$$
$$5T = 4T + 12$$
$$T = 12$$

9.

$$\frac{D}{40} \times 12 = \frac{D}{50} \times 15$$

EXERCISE 3 If, in Example 3, Anthony and Bill run in opposite directions around the track, how often will they meet?

66   *Rate Problems*

EXAMPLE 4   An airplane flew with the wind for one hour. It required 1 2/7 hours to fly back against the wind. If the cruising speed of the plane is 400 miles per hour, find the speed of the wind.

Solution
1. READ THE PROBLEM.
2. The wind speed is asked for.
3. (a) The cruising speed of the airplane is 400 miles per hour.
   (b) The distance out is the same as the distance back.
4. Here we get something: When we say the cruising speed is 400 m.p.h., we mean the plane will fly that fast in still air. But when it goes with the wind its speed is increased by the amount of the wind speed, and is reduced by that same amount when flying against the wind.
5. Let
$$R = \text{wind speed in m.p.h.}$$
6. Then
$400 + R =$ ground speed of the plane with the wind, and
$400 - R =$ ground speed of the plane against the wind
7. Then, since distance = rate × time,
$$(400 + R) \cdot 1 = \text{distance out}$$
$$(400 - R) \cdot 1\frac{2}{7} = (400 - R)\frac{9}{7} = \text{distance back}$$

By 3(b) then
$$400 + R = (400 - R)\frac{9}{7}$$

8.
$$7(400 + R) = 9(400 - R)$$
$$2800 + 7R = 3600 - 9R$$
$$16R = 800$$
$$R = 50 = \text{wind speed in m.p.h.}$$

9. We leave this for the student.

EXERCISE 4   A canoeist paddles upstream in a river for 3 hours, and back in half that time. If he can paddle five miles an hour in still water, how fast is the water flowing?

Rate Problems   67

EXAMPLE 5 It is approximately 3000 miles from Kennedy Airport in New York to Shannon Airport in Ireland. Two planes piloted by Anderson and Brown leave New York at the same time. Anderson flies 100 m.p.h. faster than Brown and arrives at Shannon one hour earlier. How fast did each fly?

Solution
1. READ THE PROBLEM.
2. The speeds of the two planes are sought.
3. (a) The distance between Kennedy and Shannon is 3000 mi.
   (b) Anderson flies 100 m.p.h. faster than Brown.
   (c) Anderson arrives 1 hr ahead of Brown.
4. Nothing here.
5. Let
$$R = \text{Anderson's speed (m.p.h.)}$$
$$T = \text{Anderson's time (hrs)}$$
Then
$$R - 100 = \text{Brown's speed (m.p.h.)}$$
$$T + 1 = \text{Brown's time (hrs)}$$
6.
$$RT = \text{Anderson's distance}$$
$$(R - 100)(T + 1) = \text{Brown's distance}$$
7.
$$RT = 3000 \qquad (1)$$
$$(R - 100)(T + 1) = 3000 \qquad (2)$$
8. From (2)
$$RT + R - 100T - 100 = 3000$$
Subtracting (1) we get
$$R - 100T - 100 = 0$$
$$R = 100T + 100$$
Substituting back into (1)
$$(100T + 100)T = 3000$$
$$T^2 + T = 30$$
$$T^2 + T - 30 = 0$$
$$(T + 6)(T - 5) = 0$$
$$T = 5 = \text{time for Anderson to fly}$$

68   Rate Problems

(What about $T = -6$?)

$$T + 1 = 6 = \text{time for Brown}$$
$$R = \frac{3000}{5} = 600 \text{ m.p.h.} = \text{Anderson's speed}$$
$$R - 100 = 500 \text{ m.p.h.} = \text{Brown's speed}$$

EXERCISE 5   Arthur is hiking up the Appalachian Trail, and Bob is hiking down it. One night they camp twenty-one miles away from each other. The next morning they both start at 8:00 A.M. and they meet at 11:00 A.M. Bob arrives at Arthur's camp one hour and 45 minutes before Arthur arrives at Bob's camp. How fast was each walking?

EXAMPLE 6   In the year 2170 a 30,000 mile long train of slow rocket freighters left for the platinum mines of Mars at a rate of 40,000 miles a day. Every five days a small survey rocket left the leading freighter, went back over the length of the train and returned to the leading freighter in one day. How far did the survey rocket travel on each trip?

Solution

1. READ THE PROBLEM.

2. The distance the survey rocket traveled on each trip is sought.

3. (a) The train's speed is 40,000 m.p.d.
   (b) The length of the train is 30,000 miles.
   (c) The round trip took one day.

4. Nothing here.

5. Let
$$R = \text{rate of survey rocket in m.p.d.}$$
Then
$$D = R \times 1 = R = \text{miles traveled in one day}$$

Rate Problems

6. (a) For the first part of the trip, the pilot of the lead freighter would see the survey rocket receding at

$$R + 40{,}000 \text{ m.p.d.}$$

and he would see it travel

$$30{,}000 \text{ miles}$$

away from him. Thus the time for that part of the trip would be

$$\frac{30{,}000}{R + 40{,}000} \text{ days}$$

When the survey rocket is coming back the freighter pilot would see it catching up at a speed of

$$R - 40{,}000 \text{ m.p.d.}$$

and again he would see it travel a distance of 30,000 miles. So the time for this part of the trip would be

$$\frac{30{,}000}{R - 40{,}000} \text{ days}$$

7. The total time is one day:

$$\frac{30{,}000}{R + 40{,}000} + \frac{30{,}000}{R - 40{,}000} = 1$$

8.

$$30{,}000(R - 40{,}000) + 30{,}000(R + 40{,}000) = R^2 - (40{,}000)^2$$
$$R^2 - 60{,}000R - (40{,}000)^2 = 0$$

By the quadratic formula

$$R = 30{,}000 + \sqrt{(30{,}000)^2 + (40{,}000)^2}$$
$$= 30{,}000 + 50{,}000$$
$$= 80{,}000 \text{ miles per day}$$

Thus the rocket travels

$$R \times 1 = 80{,}000 \text{ miles}$$

9.

$$\frac{30{,}000}{120{,}000} + \frac{30{,}000}{40{,}000} = 1$$

EXERCISE 6   If the survey rocket had taken two days for the trip, find the distance it traveled on each trip.

70   *Rate Problems*

EXAMPLE 7   Paula leaves the house each afternoon in time to pick up her husband, Paul, from work, and drive him back home. One day Paul left work forty minutes early, and started walking toward home. He met Paula on the way and they returned home arriving ten minutes earlier than usual. If the car traveled at constant speed, and no time was lost in turning around, how long did Paul walk?

Before we give the solution to this problem, we need to make some comments about it. It is a rather difficult one, probably the most difficult one discussed in this booklet. This is at least partly due to the fact that most of us attack it in the wrong way, as a rate problem. It is not really a rate problem, since *only* the times are involved. So at this point you may take this hint and look at it, concentrating solely on the relationships of the times involved.

But, like many hard problems, it is not complicated: it merely requires the right approach, and after this is seen, it may seem easy. It is these new insights and approaches, however, that present the real difficulties in mathematics, difficulties which are much harder to overcome than convoluted complications which careful analysis will untangle, even if such analysis is somewhat tedious on occasion.

Now let us turn to our formal solution of this problem.

Solution
1. READ THE PROBLEM.
2. We are asked to find how long Paul walks.
3. We write down the data. Ladies first:
   (a) Normally, Paula makes a round trip: driving to get Paul, and returning him home.
   (b) On the day in question she makes a shorter round trip, saving ten minutes.
   Now the information relating to Paul:
   (c) He leaves 40 minutes early.
   (d) He walks part way, and then rides home with Paula the rest of the way.
   (e) He arrives home ten minutes early. [From (c) and (e) it is clear that Paul spent 30 minutes more than usual on his trip home.]
4. Nothing here.

5. Let
> $x =$ time (in minutes) for Paula to drive from home to Paul's job
> $=$ time to drive back home

Let
> $y =$ time Paula drove until she met Paul walking
> $=$ time to drive home after she met Paul

Let
> $z =$ time Paul walked

6.
> $2x =$ time for normal round trip
> $2y =$ time for shorter round trip
> $y + z =$ time for Paul to get home when he walked part way
> $x =$ normal time for Paul to get home

7. From Step 6 and Step 3(b):

$$2y = 2x - 10$$

From Step 6 and Step 3(e):

$$y + z = x + 30$$

8. Dividing the first equations in Step 7 by 2 and the subtracting the result from the second equation, we get

$$z = 35$$

9. We leave this to the student.

We can really solve this problem, as we can now see, by simple arithmetic. We can argue as follows:

(1) As noted in Step 3(b), Paula saves ten minutes on the round trip.
(2) Thus she saves 5 minutes on each one-way trip.
(3) She therefore meets Paul 5 minutes early.
(4) Since Paul left 40 minutes early, and met Paula 5 minutes early, he must have walked 35 minutes.

ADDITIONAL EXERCISES

7. How soon after 3 o'clock will the hour hand and the minute hand of a clock point in the same direction?

Rate Problems

8. John starts hiking along a trail going at 4 m.p.h. Two hours later Juan starts after him on a bicycle at 12 m.p.h. How long does it take Juan to catch John?

9. At the Dogtrot County Olympics in 1870 Hank Lankman ran against Tim Thinney. Tim ran one-fifth faster than Hank around a track that measured 8 laps to the mile, and overtook him every four minutes. What were their speeds?

10. An airplane flying north at 420 m.p.h. passes over a point on the ground; 5 minutes later another plane flies west over the same point at the same speed. How long is it before they are 65 miles apart?

11. A jet plane flew 1725 miles with the wind in 3 hours. The return trip against the wind took $3\frac{12}{19}$ hours. Find the speed of the wind and the speed of the plane in still air.

12. A jet plane flew 1300 miles in two hours with the wind. On the return trip the wind was blowing in the same direction, but twice as fast as on the first trip. The return took two hours and thirty-six minutes. Find the speed of the plane in still air and the two speeds of the wind.

11
MAXIMUM AND MINIMUM PROBLEMS

EXAMPLE 1   A man has forty feet of fencing and wishes to build a rectangular pen by using an existing wall for one side and the fencing for the other three sides. How much area can he enclose, and what are the dimensions of the pen that encloses the most area?

This is a problem that one can find in one form or another in almost all calculus books, and the students learn to do such problems, and more difficult ones, dealing with finding the biggest or smallest that something can be. There is a certain class of such problems, including the one just stated, which can be solved by methods of elementary algebra.

The idea is a very simple one. It depends on the observation that if n is an *even* number then,

$$y^n$$

is never negative, whatever may be the value of y, and is zero only when $y = 0$. For example,

$$(x-3)^2$$

[i.e., $y = x - 3$, $n = 2$] is never negative and is zero only when $x = 3$.

Then

$$5 + (x-3)^2$$

is never less than 5 and is equal to 5 only when $x = 3$, and

$$16 - (x-3)^2$$

is never greater than 16 and is equal to 16 only when $x = 3$. This is true because

$$5 + (x-3)^2 = 5 + (\text{something never negative}) \geqq 5$$

and

$$16 - (x-3)^2 = 16 - (\text{something never negative}) \leqq 16$$

Now let us return to Example 1.

74   *Maximum and Minimum Problems*

Solution
1. READ THE PROBLEM.
2. The largest area of the pen, and its dimensions are asked for.
3. (a) There are 40 feet of fence to form 3 sides.
   (b) The pen is rectangular.
4. The previous comments should be clearly understood.

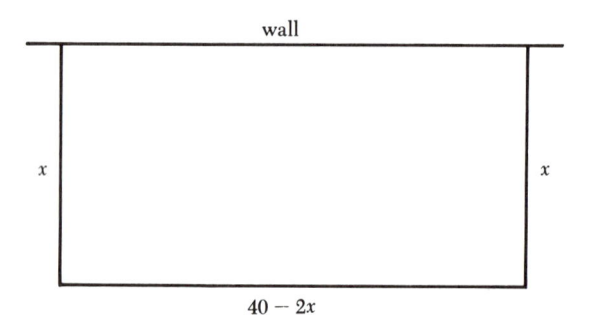

```
                    wall
         ┌─────────────────────────┐
         │                         │
     x   │                         │   x
         │                         │
         └─────────────────────────┘
                   40 − 2x
```

5. Let
$$x = \text{the length of one of the ends}$$
Then
$$40 - 2x = \text{length of a side}$$
6. The area is
$$A = (40 - 2x)x = 2x(20 - x)$$
7. & 8. This is a different sort of problem than we have looked at before and which does not lead to setting up and solving an equation. It is a question of how big the area
$$A = 2x(20 - x)$$
can be, and for which value of x it is the biggest.

We write
$$A = 40x - 2x^2$$
and complete the square on the right:
$$A = \quad -2(x^2 - 20x \quad)$$

where we have left room to fill in the required terms. We take half the coefficient of x:

$$\frac{1}{2}(-20) = -10$$

square it

$$(-10)^2 = 100$$

and add to the terms in the parentheses.

Since this amounts to subtracting 200, we compensate by adding it back in:

$$A = 200 - 2(x - 20x + 100)$$
$$A = 200 - 2(x - 10)^2$$

From this it is clear that the biggest value for A is

$$A = 200$$

which occurs when

$$x = 10 = \text{width of rectangular pen}$$

and

$$40 - 2x = 40 - 20 = 20 = \text{length}$$

EXERCISE 1 If a ball is thrown vertically into the air with an initial speed of 48 ft/sec, then t seconds later its height above ground in feet is (approximately) given by

$$h = 48t - 16t^2$$

How high will it go and when will it reach its maximum height? When will it strike the ground?

EXAMPLE 2 Find the area and the sides of the triangle of largest area having perimeter 24 ft and a base of length 6 ft.

Solution

1. READ THE PROBLEM.

2. We want the area and length of sides of the triangle with largest area.

76   *Maximum and Minimum Problems*

3. (a) The perimeter is 24 ft
   (b) The base is 6 ft

4. There is a formula (Hero's formula) for the area of a triangle:
$$A = \sqrt{s(s-a)(s-b)(s-c)}$$
where
$$s = \text{one half the perimeter}$$
$$a, b, c = \text{the lengths of the sides}$$

5.

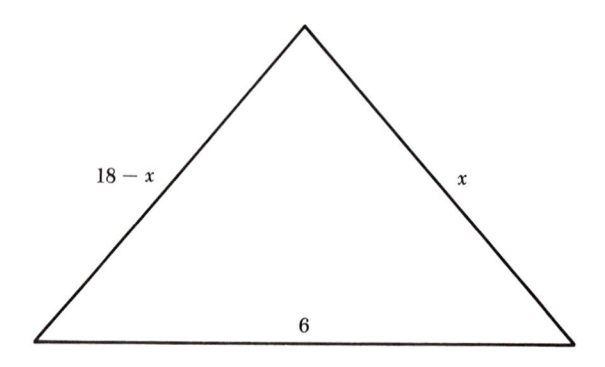

Let
$$x = \text{length of one unknown side}$$
Then
$$18 - x = \text{length of the other}$$
(Why?)

6. In applying Hero's formula
$$s = 12, \quad a = 6, \quad b = x, \quad c = 18 - x$$
$$A = \sqrt{12(12-6)(12-x)[12-(18-x)]}$$
$$= 6\sqrt{2}\sqrt{(12-x)(x-6)}$$

7. & 8. Again we merely examine the formula for the area.
$$A = 6\sqrt{2}\sqrt{\quad -72 - (x^2 - 18x \quad)}$$
$$= 6\sqrt{2}\sqrt{81 - 72 - (x^2 - 18x + 81)}$$
$$= 6\sqrt{2}\sqrt{-72 - (x^2 - 18x)}$$
$$= 6\sqrt{2}\sqrt{9 - (x-9)^2}$$

Clearly, the largest area occurs when

Maximum and Minimum Problems

$$x = 9 = \text{length of one side, and}$$
$$18 - x = 9 = \text{length of the other side}$$

So the triangle is isosceles, and its area is

$$A = 6\sqrt{2}\sqrt{9} = 18\sqrt{2}$$

EXERCISE 2 Find the dimensions and the area of the rectangle of largest area if the perimeter is 40 ft.

ADDITIONAL EXERCISES

3. A rectangle has a perimeter of 40 feet. How should the dimensions be chosen so that the diagonal is the shortest?

4. A rectangular plot is to be fenced. A cross fence, parallel to two of the sides, is to be used to divide the plot into two rectangular pens. Find the dimensions which give the maximum area of the plot if a total of 144 feet of fencing is used.

5. What is the smallest value of
$$x^2 + 2x + 15?$$

6. What is the greatest value of
$$24 - 8x - 2x^2?$$

7. What determines whether a quadratic expression $ax^2 + bx + c$ has a biggest value or a least value?

12
ACCELERATION PROBLEMS

In Chapter 10 we discussed rate or speed problems. In the beginning we talked about average rate, and for the rest of that chapter, and in all of the problems, we assumed we were dealing with constant rates, that is, for any time period, T, of the motion the distance, D, traveled in that time period was such that the ratio, D/T, was independent of the particular time period, that is, was a constant, which we called the rate, or speed, or velocity.

It is clear that in most of our real contacts with motion we are not dealing with constant rates, and when we actually use constant rates we are in fact making approximations to the real world. We will now discuss another case which is still an approximation but which does deal with a situation in which the speed is not a constant. That is, we discuss a simple case where the speed is changing. For this purpose we have to be a little careful, and so we proceed to set the background.

We will assume that the motion takes place along a straight line, and that all distances are measured from some fixed point on the line, which, as usual, we mark O and call the origin. Distances will be measured from O and will be counted positive for points to the right of O

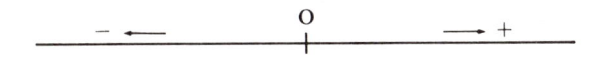

and negative for points to the left. Furthermore, if an object is moving to the right we will say that it has *positive velocity*, and if it is moving to the left that it has *negative velocity*. Thus if it is moving to the right at four ft/sec we will say its velocity is given by

$$v = 4(\text{or } +4) \text{ ft/sec}$$

but if it were moving to the left we would say its velocity is given by

$$v = -4 \text{ ft/sec}$$

80   Acceleration Problems

Speed, as distinguished from velocity, will be the absolute value of the velocity:

$$s = |v| = 4 \text{ ft/sec}$$

This fits well with our previous decision to measure distances to the right as positive and to the left as negative, for if an object has negative velocity it is moving in a negative direction and vice versa. With this setup, the velocity of an object moving on the line is the rate of change of its position, that is, the rate of change of the x which, at any given instant, denotes its position.

Suppose now an object starts from a point x_0 on our line and moves with constant velocity v. Then after time t it would have moved a distance vt and so would be at the point

$$x = x_0 + vt \qquad (1)$$

Observe that if v is negative then x is to the left of x_0, and if v is positive x is to the right. That is,

$$x < x_0 \quad \text{if } v < 0$$
$$x > x_0 \quad \text{if } v > 0$$

Now let us suppose that we no longer have a constant velocity, v, but that v is itself changing at a constant rate. That rate we will call *acceleration* and denote by a. This means that if the initial velocity is v_0 then after a time, t, the new velocity is

$$v = v_0 + at \qquad (2)$$

Again, as with v, a can be positive or negative. [Negative acceleration is sometimes called deceleration.]

Of course motions can be more complicated than those which this discussion includes, for the acceleration itself can be changing. We will not go into such questions for they are much too complicated for the mathematical level at which we are working.

We can see the significance of the sign of a by examining equation (2). Clearly, if a is negative the term at is negative and growing in size as t increases. If v_0 is positive then v is positive until the time $t = v_0/|a|$, and is negative for all larger times. This means the motion starts to the right, slows to a stop at $t = v_0/|a|$, and is then to the left for all larger t. Of course, if v_0 is negative then v simply stays negative and increases in size. In either case the velocity is negative, certainly for $t \geq v_0/|a|$, so the motion is eventually to the left. Similarly if a is positive the motion is eventually (for $t \geq |v_0|/a$) to the right.

Acceleration Problems 81

As we saw in Chapter 10, rates of speed are expressed in units of distance divided by units of time: ft/sec, miles/hr, etc. What about acceleration? This can be answered by again examining equation (2). If we solve that equation for *a*, we get

$$a = \frac{v - v_0}{t}$$

Now suppose v is in ft/sec and time is in seconds. Then the units of the numerator are in ft/sec, and the denominator in seconds. The units are then (ft/sec)/sec or ft/sec^2. We usually say "feet per second per second" in reading this. It is most often written as ft/sec/sec or ft/sec^2. Of course, other units of distance and time could be used, say miles/hr/hr or kilometers/day/day.

Now we know what we mean by constant acceleration: the velocity changes according to equation (2). It is reasonable then to ask how the position changes in this case. We know that for constant velocity, that is, for zero acceleration, the distance changes according to equation (1). But how does it change in this case? To answer this question we need to recall the remarks about average velocity which were made in the introductory paragraphs of Chapter 10.

As we said there, the distance moved in time, *t*, was the average rate times the time. Thus we have to think a little about average rate. To do this we consider the following situations. Suppose we get into a car and drive two miles down a straight road in the following ways.

I. We start out very slowly, and continue to drive very slowly, until very near the end when we speed up to 60 miles an hour.

II. We start out and quickly accelerate to 60 miles an hour and then drive at 60 miles an hour to the end.

III. We start out and continue a smooth steady acceleration to just reach 60 miles an hour at the end.

In Case I we would expect the *average velocity* to be very small, certainly less than 30 m.p.h. And in Case II we would expect the average speed to be very near 60 m.p.h., certainly greater than 30 m.p.h. But in Case III the *average velocity* should be very near 30 m.p.h., and, if we could maintain a steady acceleration, we would expect it to be exactly 30 m.p.h.

The point is this: the average velocity is not always the numerical average of the lowest velocity and the highest, but *in the case of constant acceleration* we would expect that it would be.

Therefore, if an object starts from a point x_0 on the line, with initial velocity v_0, and moves with constant acceleration a, then after time t, the velocity is

$$v = v_0 + at$$

The *average velocity* is

$$\frac{v + v_0}{2} = \frac{(v_0 + at) + v_0}{2} = v_0 + \frac{1}{2}at$$

so the distance *traveled* is

$$d = (\text{av. veloc.}) \times \text{time}$$
$$= \left(v_0 + \frac{1}{2}at\right)t$$
$$= v_0 t + \frac{1}{2}at^2$$

Its *position* x on the line is then the original x_0 *plus* the distance traveled:

$$x = x_0 + v_0 t + \frac{1}{2}at^2 \tag{3}$$

The most important elementary example of constant acceleration is that of falling bodies. If an object is falling near the surface of the earth and if it is sufficiently compact and falls a relatively short distance, then air resistance does not significantly influence the motion. (Clearly a feather does not fall the same as a buckshot pellet. This difference is due to air resistance.)

Physicists tell us that a body near the surface of the earth undergoes a gravitational pull. When the object is at rest, this is what gives it weight. When it is freely falling with negligible air resistance, gravity imparts to it an acceleration of about 32 ft/sec/sec. This information is the result of physical experiments. We can now take this information and investigate some of its mathematical consequences. In all that follows we will assume (a) air resistance is negligible and (b) the acceleration due to gravity is 32 ft/sec/sec and that it is directed vertically downward.

EXAMPLE 1 A ball is thrown straight up from the ground level with an initial speed of 48 ft/sec. (a) How high does it go? (b) How long does it take to reach the highest point? (c) How long before it returns to strike the ground? (d) With what speed does it strike the ground?

Acceleration Problems   83

Solution
1. READ THE PROBLEM.

2. The maximum height of the ball, the time necessary to reach the maximum height and to return to the ground, and the speed at striking the ground are asked for.

3. (a) The ball is thrown up.
   (b) The initial velocity is 48 ft/sec.

4. From the previous discussion we assume the acceleration is 32 ft/sec/sec in a downward direction.

5. Let
   t = time (in seconds) from when the ball was thrown
   v = velocity of ball
   a = acceleration of ball
   h = height of ball

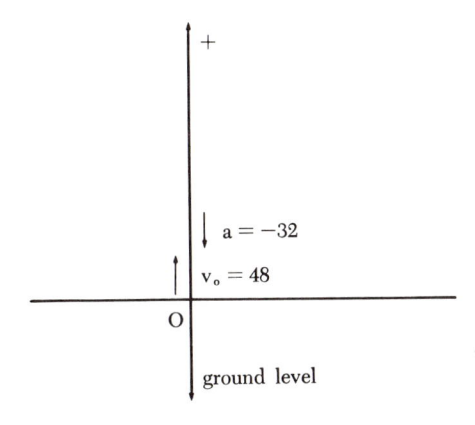

6, 7, 8. Since the ball is thrown straight up, it will move on a vertical line. Thus take this as our coordinate line, and take the origin of the line to be at ground level. We can now choose the positive direction on this line to be either up or down. Let us choose it to be up. Then we have

$$a = -32 \quad \text{(why?)}$$
$$v_0 = +48$$
$$x_0 = 0$$

Acceleration Problems

$$h = 48t - 16t^2$$
$$= -16[t^2 - 3t]$$
$$= 36 - 16\left[t^2 - 3t + \frac{9}{4}\right]$$
$$= 36 - 16\left(t - \frac{3}{2}\right)^2$$

Clearly the greatest value for h is 36 when $t = \frac{3}{2}$.
This answers parts (a) and (b) of the problem.

To go on to the next part we observe that when the ball strikes the ground, h will be zero. Then

$$0 = 48t - 16t^2$$

from which

$$t = 0 \quad \text{or} \quad t = 3$$

The time $t = 0$ was when the ball was thrown, so it strikes the ground 3 seconds later. The velocity is given by

$$v = v_0 + at = 48 - 32t$$

at $t = 3$, we have

$$v = 48 - 32 \cdot 3 = 48 - 96$$
$$= -48$$

Thus the ball returns with the same *speed* with which it was thrown, and the minus sign reflects the change in *direction* of the velocity.

EXERCISE 1 Do this problem by choosing the positive direction downward.

EXAMPLE 2 From the top of a tower 144 feet high a ball is thrown vertically upward with an initial velocity of 40 ft/sec. (a) How high does it go? (b) When will it strike the ground?

Solution
1. READ THE PROBLEM.
2. We are asked to find
   (a) the maximum height of the ball.
   (b) the time to strike the ground.

3. (a) The ball is thrown upward.
   (b) The initial velocity is 40 ft/sec.

4. The acceleration is 32 ft/sec/sec downward.

5. Using the same notation as in the previous example, but using

$$h_0 = 144$$
$$v_0 = 40$$

we get

$$h = 144 + 40t - 16t^2$$

6, 7, 8. Then, completing the square,

$$h = 144 + 25 - 16\left(t^2 - \frac{5}{2}t + \frac{25}{16}\right)$$

$$h = 169 - 16\left(t - \frac{5}{4}\right)^2$$

From this we see that the largest h is

$$h = 169 \text{ ft}$$

at

$$t = \frac{5}{4} \text{ sec}$$

which solves part (a).

The ball strikes the ground when $h = 0$, that is, when

$$16\left(t - \frac{5}{4}\right)^2 = 169$$

$$4\left(t - \frac{5}{4}\right) = \pm 13$$

$$t - \frac{5}{4} = \pm \frac{13}{4}$$

$$t = \frac{5}{4} \pm \frac{13}{4}$$

Thus

$$t = \frac{9}{2}$$

or

$$t = -2$$

so the ball strikes after $9/2 = 4\frac{1}{2}$ sec. What physical significance can we give to the "extraneous root" $t = -2$?

With what velocity does the ball strike the ground?

Acceleration Problems

EXERCISE 2   A ball is thrown vertically downward with an initial velocity of 72 ft/sec from a height of 88 ft. When and with what velocity will it strike the ground?

ADDITIONAL EXERCISES

3. A cowboy boasted that he could drop a washer from his shoulder, reach for his gun and shoot a bullet through the hole in the washer as it passed the level of his gun. If the washer fell two and a half feet, how much time did he have after dropping the washer?

4. A race driver is going down the track at 60 m.p.h. and slams the brakes on. If the brakes apply a constant deceleration (i.e., negative acceleration) and if he stops in 100 yards, how long does it take him to stop?

5. A stone is dropped from the top of a high building to a concrete walk at the base of the building. The sound of the stone striking is heard 9 seconds after the stone is dropped. If sound travels at 1024 ft/sec, how high is the building?

6. Find a general formula for the height of the building in problem 5, in terms of the time, T, required to hear the sound of the striking stone.

ANSWERS

CHAPTER 4

1. 12, 36
2. 10, 37
3. 10, 15, 18
4. 8, 10, 16
5. 5, 30
6. 4 or 6
7. 70
8. 4, 5, 30

CHAPTER 5

1. 26
2. 165
3. 8 half dollars, 11 quarters, 6 dimes
4. 20 nickels, 40 dimes
5. 50 nickels, 50 dimes
6. 74
7. 74
8. 30, 31
9. 7, 12
10. 5, 6

CHAPTER 6

2. 3, 6
3. 70, 74
4. $r = 12$ cm, $h = 64$ cm
5. 4, 5, 12
6. $8\frac{1}{2}$, $5\frac{1}{2}$
7. $5\frac{1}{2} \times 8$ or $6 \times 7\frac{1}{3}$
8. 12×32
9. 5×10
10. 11, 12

CHAPTER 7

1. 15
2. 3
3. 6 each
4. 4, 6
5. $82125 \div 140 = 857$ (approx.)

CHAPTER 8

1. 3mg 72%, 5mg 84.8%
2. (a) 26, (b) $14\frac{4}{9}$
3. $\frac{3}{5}$ ($2.00); $\frac{2}{5}$ (75¢)
4. 30g (5%); 150g (25%); 50g (60%)
5. 7.5
6. $1\frac{3}{5}$ (75¢); $1\frac{2}{5}$ ($1.00)
7. $2\frac{2}{7}$
8. 12, 18

CHAPTER 9

1. $6000 (6%); $4000 (4%)
2. 5%
3. $5,000; $10,000

CHAPTER 10

1. 4, 12
2. $65\frac{5}{11}$ min
3. $22\frac{2}{9}$ sec
4. $1\frac{2}{3}$ m.p.h.
5. 3, 4
6. $(3 + \sqrt{73})10,000 = 115,000$ (approx.)
7. $16\frac{1}{16}$ sec
8. 1 hr
9. $11\frac{1}{4}$, $9\frac{3}{8}$ m.p.h.
10. $3\frac{4}{7}$ min
11. 50; 525 m.p.h.
12. 50; 600 m.p.h.

CHAPTER 11

1. 36 ft, $1\frac{1}{2}$ sec; 3 sec
2. 10 × 10; 100
3. 10 × 10
4. 24 × 36; 864
5. 4
6. 32
7. max if $a < 0$, min if $a > 0$

CHAPTER 12

2. 1 sec, 104 ft/sec
3. $\sqrt{10/8} = 0.4$ sec (approx.)
4. $6\frac{9}{11}$ sec
5. 1024 ft
6. $1024[32 + T - 8\sqrt{16 + T}]$